George E Waring

How to Drain a House

Practical Information for Householders

George E Waring

How to Drain a House
Practical Information for Householders

ISBN/EAN: 9783744678650

Printed in Europe, USA, Canada, Australia, Japan

Cover: Foto ©berggeist007 / pixelio.de

More available books at **www.hansebooks.com**

How to Drain a House

PRACTICAL INFORMATION FOR
HOUSEHOLDERS

BY

GEO. E. WARING, JR., M. INST. C. E.
Consulting Engineer for Sanitary Drainage.

SECOND EDITION, WITH ANNOTATIONS

NEW YORK

D. VAN NOSTRAND COMPANY
1895

CONTENTS.

ILLUSTRATIONS.

NOTE.—The text of this book applies to conditions existing in 1885. The notes at the ends of the chapters relate to progress made since that date.

PREFATORY: OUR ENEMY THE DRAINS.

The drains in average modern houses are probably the most serious and prevalent enemies with which struggling humanity has to contend. But, at the worst, they are only incidentally enemies, and they are *never* necessarily so.

All of the intended purposes of the drains are wholly beneficial. With all of their defects, it is not too much to say that on the whole the world is much better off with them than it would have been without them. Defective though they often are, whether in the house or in the street, they have probably been, next after the introduction of a pure water-supply, the most important factor in the reduction of the death-rate.

That a country town, depending entirely on outside privies and on cesspools,—with all that their use implies in the way of exposure and irregularity of habits, and of the fouling of the ground and the air—would be vastly improved in its sanitary condition by the introduction of even an imperfect system of sewerage and with plumbing-work of a very low and ordinary character, is not to be doubted. Therefore, at the worst, one enemy has displaced another, and the new one is much less to be dreaded than the old.

What we still need is continued progress in the right direction. As indoor water-closets are better than out-of-door privies ; as defective waste-pipes are better than none ; and as bad sewers are better than cess-pools, so will good substitutes for all of these defective things lead to still further improvement.

In the following pages the subject is approached entirely from the point of view of the individual householder—he who has got so far on in wisdom as to know that imperfect drainage is an enemy to the well-being of his household, and that by abolishing the imperfections, the enemy can be disarmed and made a most useful ally.

The drainage system is, however, a trustworthy ally *only so long as the woman of the house* holds it under close and careful supervision.

Her whole duty is not done when her husband has paid a good round sum to the engineer and to the plumber. It is only begun.

There has been placed under her control a means of safety, or an engine of destruction, according as she performs her duty, or neglects it. She can not safely delegate her responsibility to her servants. Her own eye must see that at no point, has neglect, at any time, permitted even the beginning of filth—for the beginning of filth is the beginning of danger. It marks the desertion of the ally to the ranks of the enemy.

G. E. W., Jr.

NEWPORT, R. I., December, 1884.

HOW TO DRAIN A HOUSE.

CHAPTER I.

HOUSE DRAINS AND HEALTH.

FORTY years ago the best houses in American cities were little if at all better in the matter of drainage than are the best houses of Paris, with very rare exceptions, to-day. They had at best only one drain to remove the kitchen waste and another to drain the cellar. All the water used was carried by hand and it was used in limited quantities. The bath was an exceptional luxury ; the water-closet was almost unknown, and the unspeakable horrors of the privy, the close-

stool and the sick-chair—still dominant in
undrained houses, and especially in country houses—were accepted as an inevitable
incident of human life. Happily, we are
now emerging from this barbaric condition,
and are learning to regulate our appliances
according to the dictates of health and
decency. A dozen years ago in a pamphlet
description of the earth-closet, then recently invented, I wrote the following,
which is as true of country houses now
as it was then :

Out-of-door privies, those temples of de-
fame and graves of decency, that disfigure
almost every country home in America,
and raise their suggestive heads above the
garden-walls of elegant town-houses, are, I
believe, doomed to disappear from off the
face of the earth. Thirty years ago, every
back-yard in New York City was provided
with one of these buildings ; now, since
the water-closet has come into universal

use, probably there are not twenty of
them to the square mile. Twenty years
hence, it is to be hoped, they will become
equally rare in smaller towns and in the
country. That they are objectionable
on the score of decency and comfort, will
be confessed by all. What is not so generally understood is their pernicious
effect upon health. The influence of
subterranean stores of fecal matter in
the propagation of disease has already
been referred to, and will be more fully
discussed hereafter ; but that which pro-
duces, in the aggregate, far worse results—the aggravation of the difficulties of
delicate females—has attracted less attention than its importance deserves. It is
universally admitted that nothing is more
injurious to health than irregularity and
the undue retention of the rejectamenta
of the intestines. It is not necessary to
quote scientific authority to prove to any

person of intelligence that in prompt and regular attention to this duty lies the cardinal secret of health. We have all been reminded, in our own persons, that our health and efficiency, as well as our cheerfulness and good humor, depend on perfect regularity in this regard. There can be little question that the prevailing female complaints are often induced, and always intensified, by disorders of the digestive organs, and the oppression in the lower regions that neglect in this matter causes. Admitting the justness of the view, let us see what chance a woman living in the country has to escape the direst evils that "delicate health" has in store for its victims. The privy stands, perhaps, at the bottom of the garden, fifty yards from the house, approached by a walk bordered by long grass, which is always wet except during the sunny part of the day, overhung by shrubbery and vines,

which are often dripping with wet, and
sometimes exposed to the public gaze. In
winter, snow-drifts block the way, and during rain there is no shelter from any side.
The house itself is fearfully cold, if not
drifted half-full with snow or flooded with
rain. A woman who is comfortably
housed during stormy weather will, if
it is possible, postpone for days together
the dreadful necessity for exposure that
such conditions imply. If the walk
is exposed to a neighboring work-shop
window, the visit will probably be put off
until dusk. In either case, no amount of
reasoning will convince a woman that it is
her duty, for the sake of preventing
troubles of which she is yet ignorant, to
expose herself to the danger, the discom-
fort, and the annoyance that regularity
under such circumstances implies. I
pass over now the barbarous foulness
and the stifling odor of the privy-

vault. It is only as an unavoidable evil that these have been tolerated; but I can not too strongly urge attention to the point taken above, and insist on the fact that every consideration of humanity, and of the welfare not only of our own families, but of the whole community, demands a speedy reform of this abuse. It will hardly be believed by my more civilized readers that, over more than half of the older settled parts of the United States, even the every-way objectionable system that I have described is comparatively unknown, and that the corn-field and the thicket are the only retreat provided, while the majority of farmers' houses, even at the North, are most inadequately supplied. In view of the foregoing facts, I make no apology for calling the attention of women themselves to this important matter, believing that they will universally concede that, however much of elegance and comfort may surround them in the appointments of their homes, their mode of life is neither decent, civilized, nor safe, unless they are provided with the conveniences that the water-closet and the earth-closet alone make possible.

As a positive source of disease, and as the occasion of a most injurious irregularity, the barbarous appliances of our ancestors, still existing in connection with nine-tenths of the habitations of the United States, were and are doubtless more injurious, even at the arm's length at which they were held, than are the average water-closets of average city houses. In saying this, however, it is not intended to be understood that these average modern appliances are acceptable as any thing but makeshifts—though relatively good, they are absolutely bad.

That their injurious effect on health is

practically as bad as theoretically it ought
to be, is not obviously true. Many of
their victims die in infancy, and so large
a number of those who pass this critical
period withstand their evil effects, that it
has come to be believed by the people at
large that the outcry against them is an
unreasonable one.

Perhaps all popular outcry is unreasonable, but certainly those who will take the
trouble to investigate the condition of the
drainage of an average house, supplied
with the usual plumbing appliances, will
find defects at every turn—not merely
slight defects which it would on the whole
be better to avoid, but generally very
grave defects which it is absolutely necessary to eradicate before we can hope to
secure those conditions of perfect health
which we have a right to demand of the
civilization of which we boast. In periods
of epidemic, or when cholera or yellow

fever is apprehended, the popular imagi-
nation on the subject becomes excited,
and the long death-roll which pestilence
creates gives an almost dramatic force to
the stronger arguments advanced against
our imperfect plumbing work.

As a matter of statistics, however, the
deaths caused by any epidemic, and the
degree to which these are favored by bad
drainage, are of very secondary importance. The thousands of deaths from
yellow fever in New Orleans, and in
Memphis, and in the Mississippi Valley
generally, in 1878 and 1879, fairly shook
the country with terror. They amounted
in all, in both years, to less than twenty
thousand.

Their suddenness and their concentration gave them their striking effect.

In the country at large there are annu-
ally not fewer than one million deaths, and
not fewer than two hundred thousand of

these are from directly preventible diseases. Not fewer than one hundred
thousand of these latter probably owe
their origin to diseases occasioned by defective drainage or by the improper retention of fecal matter and other organic
wastes.

This enormous preventible death is,
*from the point of view of the political
economist,* only an index of something
worse. Each preventible death doubtless
represents, taking one disease with another, twenty cases of preventible sickness, and each such case of sickness
implies at least twenty days of suffering
and disability with its serious incidental
cost in nursing and medication.

The real benefit, therefore, that is to
accrue to the community from the establishment of perfect sanitary regulations,
in the house and in the town, aside from
the establishment of greater vigor and

efficiency, and of increased ability to withstand insalubrious conditions, is to be sought not so much in the prevention of these deaths as in the abolition of the diseases which cause them.

As in the town, so in the individual house, we shall be safe if our attention is given only incidentally to the saving of life, but directly to the preservation of health, *i. e.*, to the removal of all those conditions which affect the purity of the atmosphere in which we live, involving, of course, the purity of the ground on which our houses are built and the absolute prevention of the putrefaction any where within or near the house of its organic offscourings.

NOTE.—In all of the larger towns there has been, since 1885, a very marked improvement in the character of the plumbing work done. In the older town houses, and in village and country houses, little advance has been made.

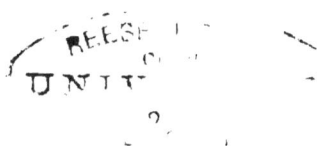

CHAPTER II.

A HOUSE completely equipped for convenient modern life has two systems, comparable to the arterial and venous systems of the living animal. Its water supply is taken from the street main, or from the private reservoir or tank, and carried through tight pipes to the different points where convenience requires it to be delivered. This supply may easily be secured against contamination, and there is little trouble, if it is pure at its source, in keeping it pure until it is delivered for use. The venous system—that which has for its office the removal of the water after use—the

water supply *plus* the burden of filth
which has been added to it—has its main
discharge through the outlet drain.

The various points where the pure
water is made foul are also conveniently
distributed throughout the house. The
drains begin usually as small lead pipes ;
these lead to larger iron pipes and these
to the main channel, sometimes of iron and
sometimes of earthenware. The particular
office which this venous system has to per-
form is to remove all of the refuse of the
house which is suited for transportation
in water. This office it usually performs
with such a degree of completeness as to
satisfy the average demand of the householder, *i. e.*, when water is discharged
from a vessel it runs into the waste-pipe
and is not heard from again. What
becomes of it after it passes out of sight,
whether or not it and its filth are really
removed from the house ; whether or not

the removal is so rapid and complete as to prevent deposits and accumulations, are questions about which the average member of the community is heedless.

Lead pipes are usually tight. Iron pipes are connected together funnel-wise, so that whether they are tight or not the water runs away, and if the main drain does not carry to the outlet all that it receives, it leaks it out into the ground, where it is out of sight.

There is more or less unpleasant smell here and there and often a sensation of closeness about the atmosphere of the whole house. These attract little attention, because foul smells and closeness are generally accepted as a necessary incident of the production and disposal of filth.

It is precisely the conditions which lead to these more or less obviously disagreeable sensations which constitute the main feature of the practical sanitary question.

I have now in mind a large, elaborately decorated and richly furnished
house in Fifth Avenue, New York, which
I was called to inspect after a serious
case of illness. The owner believed its
drainage works to be all right. They
had cost a very large sum originally and
they had been repaired at considerable
outlay.

Stationary wash-basins were distributed throughout the house. The spaces
under them were closely encased in
fine cabinet work. There were two bath-
rooms on each floor, containing the
usual assortment of fixtures. The water-
closets — their bowls elaborately decorated, some of them with gilt—were the
usual " pan " closets of the period. Every
pipe was concealed behind partitions, and
all traps and machinery were covered
under the floor or behind elaborate carpentry. The kitchen and pantry sinks

had waste-pipes two and a half inches in diameter, but were so filled with grease as to have but a thread of water-way left. The drains ramified in various directions under the floor of the cellar. These carried not only the wastes of the house, but the water from the roof. The waste-pipes in the house, in compliance with what was then believed to be the best practice, were vented with small air-pipes leading above the roof. There was no effective circulation of air in any part of the drainage system. There were evidences of positive leakage at many points, and the lines occupied by the supply and waste-pipes continued, in groups, through every floor, so that the whole house, behind the partitions and between the floor beams, constituted an unobstructed air channel from the ceiling to the garret. A stale odor was perceptible in all the passage-ways and closets and staircases,—

wherever the open fire places did not afford free ventilation.

It was demonstrated in the course of the inspection and renewal of the work, that the matters intended to be removed from the house were very largely only removed from sight. For lack of any thing like adequate flushing, the large waste-pipes and soil-pipes ·were lined with offensive slime; the kitchen and pantry sink wastes were nearly filled with putrefying food involved in the congealed grease that had attached itself to their walls. The drains under the cellar leaked at many of their joints so as to saturate and render foul the earth under the concrete floor. A direct opening into the sewer, intended to drain the cellar floor, protected only by a bell trap, from which the water had evaporated, was pouring into the air of the house as much foul gas from the drain as could get past the accumulated dust and cobwebs by which it was nearly obstructed.

The chief complaint of the owner, and what he believed to be the source of all his woe, was that the public sewer was a foul one and that his house was pervaded with sewer gas. This may or may not have been the case. Whether it was or not, it was clearly demonstrated that there existed in the house itself, and almost throughout its whole drainage system, from the "container" of every water-closet, coated on the inside nearly half an inch thick with fecal matter, to the point where the main drain passed through the foundation wall—a more than ample source of all the difficulty.

Certain it is that when the work was reconstructed, though the sewer remained as before, the atmosphere of the house became, and still continues, perfectly pure.

The reconstruction of the work was substantially in accordance with the principles set forth in the following pages.

This is certainly not an exaggerated illustration of the condition of the drainage works of even the best houses of only a few years ago. Since then, not only has workmanship been greatly improved, but a number of radical modifications and reformations have been universally adopted. In some cities these reforms are enforced by rigid regulation of the health authorities. Nevertheless, a great majority of the houses in New York city as well as elsewhere are still in a very defective condition, and certain requirements, which the better sanitary practice considers important, are almost entirely absent.

The glaring mistakes of ten years ago are no longer repeated by reputable plumbers. The minor and more modern improvements which are still less generally appreciated will doubtless, before very long, be adopted into general practice, and the whole community will benefit greatly thereby.

It is still to be urged on every man and woman, who realizes the importance of perfection in this vital element of house construction, to anticipate the universal acceptance of the better processes and to insist that in their own cases at least, they shall be adopted forthwith.

The apprehension of cholera may stimulate attention to the subject, and that it may be urged on by each localized outbreak of typhoid fever or other zymotic disease is to be expected ; but much more than this is to be desired and to be advised, *i. e.*, to put the whole house into such perfect sanitary condition as to its waste-pipes, and as to its drains, that it can no longer by any possibility be a source even of the malaise, headache, dullness, neuralgic affection, etc., which rob life of so much of its comfort and usefulness. One may properly recall in this connection Poor Richard's recommendation : Take care of the pence and the pounds will take care of themselves.

Look out well for the health-rate ana the death-rate will lose its significance.

CHAPTER III.

THE following chapters are presented as a simple and direct statement of some positive knowledge, and of more confident belief, about the drainage of houses.

They are not addressed to that indifferent public which sees a good deal of nonsense in the theories of all reformers. They are not addressed to plumbers, who, as a rule, are little attracted and less influenced by what is said by any body whose working years have not been given to plumbing work. They are not even addressed to architects and engineers, who, whatever their own convictions, when they have convictions on this subject, so often find it necessary to compromise with their mechanics and with their clients, and to be content with such improvements as it seems under the circumstances judicious to insist on. They are addressed to that limited class who are willing to learn, and with whom a promising suggestion becomes a fruitful germ ; to the few who will agree with their teachings, and to the more who will take their propositions into earnest consideration without the intention, and often without the result, of agreeing with them.

Where they can be avoided, alternative suggestions will not be made. If there are two ways of doing a thing, one right and the other only not wrong, the right way alone will be described. There is usually but one best way, and all that is to be considered here is purely and simply the best way of improving the drainage of a human habitation, and of maintaining its good sanitary condition.

CHAPTER IV.

THE house, and the ground under and about it, and the air with which it is filled and surrounded, should be as dry and as clean as the best constant effort can make them. To this end, the most intelligent care and the most earnest attention must be given to all details of construction, and, no less, to the details of maintenance. No house, however perfect its original condition, can remain in perfect condition if subjected to the deteriorating influences of even ordinary carelessness.

Many a palace is a pig-pen in its hidden recesses, and where the light of day and the eye of a scrupulous housekeeper are withheld, there will those enemies of the

human race, dirt and damp and decay,
surely make their stand. The whole range
of cubby-holes, dark cellars, uninspected
closets, and those spaces about pipes and
fixtures which are screened from observation and withdrawn from the reach of care
by the pernicious carpentry to which the
plumbing art is so closely wedded, are, all
of them, places to be suspected and as far
as possible to be abolished. Where dark
places must be maintained, they should
be the chief objects of the householder's
care. It is a wise old sanitary saying that
" where daylight can not enter the doctor
must."

Houses that are perfect, even in the
general arrangement and construction of
their drainage works, are extremely rare.
Those which, having begun perfect, continue so under daily occupation, are still
more rare. So true is this that it is sometimes asked if it is, after all, worth while

to encounter the additional expense and the constant attention that perfection demands; whether, indeed, the world has not got on so well in spite of grave sanitary defects that it is futile to hope for an improvement corresponding with the cost in money and time.

The most simple and the sufficient answer to this is that the world has not got on well at all, and is not getting on well; that among large classes of the population one-half of all the children born die before they attain the age of five years; that those who come to maturity rarely escape the suffering, loss of time, and incidental expense of unnecessary sickness; that the average age of all mankind at death is not one-half of what it would be were we living under perfect sanitary conditions; that one of the chief items of cost in carrying on the world, to say nothing of the cost of bury-

ing those who die, is that of supporting
and attending the sick and helpless ; that
another great item is the cost of raising
children to, or toward, the useful age, and
then having them die before they begin
to make a return on the investment ; that
the great object of a well-regulated life is
to secure happiness for one's self and for
one's dependents, an aim which is crushed
to the earth with every death of wife or
child or friend.

There is a sentimental view, no less
important, which need not be recited, but
which is sufficiently suggested to the
minds of all who have had to do with the
sanitary regulation of houses by the frequency with which their services are
called into requisition only when the
offices of the undertaker have been performed.   No cost and no care would be
too great to prevent the constantly recur-
ring domestic calamities which have had

their origin, and which have found their development, in material conditions that a little original outlay and a constant and watchful care would have prevented.

The objects to be attained in the drainage of a house and of its site are, first, to remove all causes of excessive dampness ; and, second, to provide a means for the water transportation of organic wastes to a safe point of disposal, in such a way as to prevent decomposition on the premises, and so as to exclude from the house all air which has been in contact with these matters after their discharge into the drainage system.

The means for accomplishing these ends are of two distinct sorts : one allied to the drainage of agricultural lands, the other to the flushing of gutters.

CHAPTER V.

THE first in order of execution, and although not first in importance, still of absolute importance, is the work for preventing undue dampness of the interior atmosphere, or of the walls, of the house by an actual inflow of water, by an exhalation of vapor from the water contained in the soil, or by a soaking of the foundation. In the case of city houses occupying the whole width of the lots on which they stand, this drainage is necessarily confined to the cellar and foundations, and, as a rule, the water to be drained away can be delivered only into a public sewer—though there are frequent

exceptional cases where, by piercing an
impervious stratum of clay or other
material, an outlet may be gained into a
porous stratum of gravel or sand below.

Wherever the site is on a deep and naturally well-drained bed of sand or gravel,
the question of drainage as a means for
removing soil-water does not present
itself. But here another very serious
difficulty is to be encountered, having a
different sanitary bearing, but of no less
sanitary consequence. This relates to
the protection of the house against exhalations from the ground—not of moisture,
but of the atmospheric impurities of the
subsoil.

In the case of a country house, or of a
town house standing in the center of a
considerable area, it is often the most
efficient means for securing satisfactory
drainage to apply a very thorough system
of underdraining to the whole area about

it and for some distance away, by laying independent lines of tile drains, not necessarily under the house at all, but so as to surround it on all sides from which water flows toward it, and in all cases at a depth considerably below the level of the cellar-bottom.

It is seldom, even where a spring is struck in digging the cellar, that such drains, surrounding the site of the house, will not entirely divert the water. In this drainage of large lots, the character of the outlet is of secondary importance. All that is needed is that it shall be low enough for the free discharge of the flow of the drains. If the discharge be into a sewer, the drains should descend toward it with a sufficient fall to prevent foul water from setting back into them in the case of a gorging of the sewer at a point near the house.

In the drainage of a city house occupying the whole width of the lot, the same system is to be adopted, save that the drains, instead of being so placed as to surround the house and cut off water approaching it, must perforce be placed under or near the foundation to receive such water as may have reached its actual site. Here the question of outlet becomes a serious one. If the discharge must be into a sewer, then some special means must be adopted for preventing the return of the air of the sewer to the subsoil under the house.

In the construction of these drains two courses may be pursued with perhaps an equally good result. One is, after having excavated the ditch and cleared its bottom of all loose dirt, to fill in to the depth of a foot with sand or gravel— and even fine sand will answer the purpose. The other is, to use agricultural drain-tiles, preferably of the smallest

size, say an inch and a quarter in diameter, laid at the bottom of a well-graded
trench and continued to the point of
outlet.

Where tiles are used, the joints should

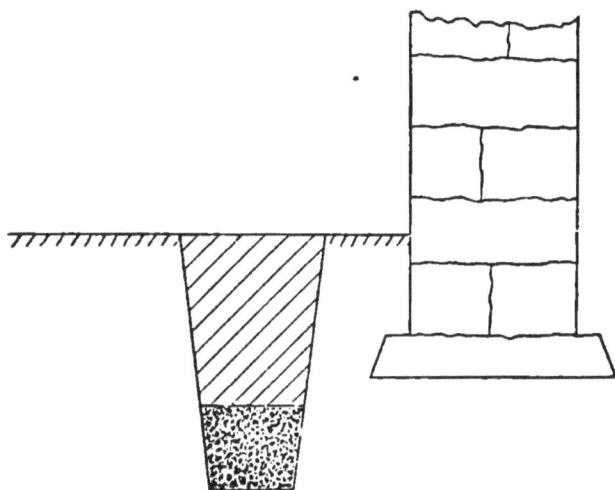

FIG. I.—GRAVEL DRAIN, UNDER CELLAR FLOOR, NEAR FOUNDATION.

be wrapped twice around with strips
of muslin drawn tight. This makes
a perfect collar, holding the tiles in
line, and affording much the best protection that has yet been devised against
the ingress of sand or silt, which usually

finds its entrance at the lower part of the joint, flowing in with the water as it rises with the general water-level and flows off over the floor of the tile.* Before the muslin will have rotted away the soil will become so compacted as not to follow the water into the tiles.

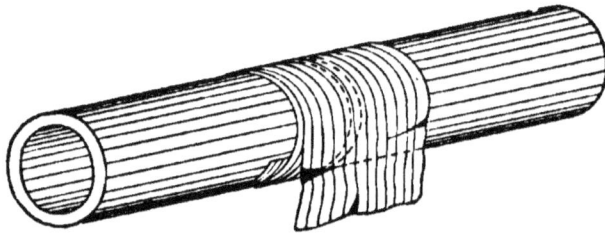

FIG. 2.—THE DRAIN, WITH MUSLIN JOINT.

Where tile drains are used, it is a mistake to marry them to other materials. Tile alone or gravel alone will make a very good drain—tile and gravel together, not nearly so good when per-

---

* This use of muslin is patented, but it is hereby dedicated to the public to the extent of its use under or within the foundation-wall of bui'dings.

manence is considered. Tiles should be
laid on the bottom of a perfectly graded
ditch, and should be compactly imbedded
in the heaviest loam that is found in
excavating. When covered to the depth
of a foot, this clay should be well trodden
down, so that if the tile could be taken
out, leaving the earth undisturbed, we
should find a complete matrix, or nidus,
which had clasped it firmly at every
point. The old marvel, How gets the
water in ? is too long for discussion here.
I beg the reader to take the word of an
old drainer that it does get in—and get
out—perfectly.

The large pipe drains with wide joints,
often with fractures giving access to vermin—no less than the "box drains,"
"French drains," "blind drains," and
various other antique devices for getting
rid of soil-water — are costly, cumber-
some, and in the long run, inefficient,

owing to their liability to obstruction.
The amount of water that can ever be
collected as a constant stream, except in
the case of a very copious spring, even
in very wet foundations, is extremely
slight. A sand seam in the natural soil
one-fourth of an inch thick is generally
sufficient to carry it ; and it is such
seams, carrying water in a slow but constant ooze, which usually produce our
subterranean and surface springs.

A tile an inch and a quarter in diameter will carry more water than can often
be collected for a constant flow from
the subsoil of half an acre of ground. A
body of sand or gravel ten or twelve
inches wide and of equal depth can not be
so compacted, provided clay and loam be
kept out of it, that it will not afford a free
outlet for all the water that can reach it
under these circumstances from the soil
of an ordinary town lot.

As a rule, the tile will be found to
be much cheaper than the other material. It is better always that the depth
of the drain should not be less than
two feet below the level of the foot
of the foundation. The more rapid
the descent the better, but even two
inches in a hundred feet, with perfect
grading, will remove a very large flow.
Indeed, if the drain has no fall, or even
if it be depressed in places, provided it
have a good and unobstructed outlet and
well-protected joints, its surplus water will
be discharged as soon as the general level
of the water reaches the overflow point.

Where the water is to be delivered to a
sewer, I should in any case recommend
the making of the outlet drain, or a part
of it, with sand or very fine gravel. I
should at least make a break ten feet long
in the course of the drain, and fill this
with such material—fine enough not to

allow the free transmission of sewer air to
the drains under the house, which a continuous tile drain would permit. I am
aware that this recommendation is radi-
cally different from what has generally
been set forth ; but it long ago com-
mended itself to my judgment, and has
proven in practice 'to be entirely successful.

It is a usual custom to connect the un-
derdrains of a house with the drain carrying the foul water, and to connect with
them, also, the rain-water conductors from
the roof. In view of what we know of
the ease with which the contained air of
the subsoil may be contaminated, it is of
the utmost importance, where the best
results are sought, to deliver the underground water itself by an independent
line guarded with absolute completeness
against the possible invasion of sewage or
foul air. Nowhere within the house, nor,

indeed, for some distance outside of it, should even the rain-water conductors deliver directly into this system.

By the means just described, the actual superabundant water of the soil may be removed.

In connection with the foundation and cellar, two things else demand attention. The first is the carrying up of dampness through the foundations into the walls of the house, and the exhalation of watery vapor, which, in the case of a heavy soil, however well drained, is of considerable amount. These difficulties attach chiefly to clayey ground. The next is the entrance into the house of the aerial exhalations of the soil.

Even a clay soil contains a large amount of air, and under different circumstances, such as changing barometric pressure, the rise and fall of water in the soil, and the action of winds, producing a strong draught in chimneys, this air enters the cellar and the house. This difficulty, not serious in the case of stiff clay, increases greatly as the soil grows more porous and becomes more dry. For example :

A pile of stones broken to the size of road-metal contains a very large amount of air,—how large we could determine by filling the voids with water and measuring its quantity. Every wind that blows, every change of temperature, every rise of water into the mass, drives out or changes a portion of this air. If at the bottom of the heap there lay a mass of carrion, its stench would be almost as perceptible as though the stones were not there. A bed of such stones sufficiently large and sufficiently compacted would make a dry, firm, safe foundation for a house—in many respects an excellent foundation.

But if the atmosphere of the house were
not separated from that of the interior of
the mass of stones by something much
more effective than even the usual cellar-
bottom concrete. and if the carrion were
putrefying beneath, the state of things
would not be the worst possible only because the obvious offensiveness resulting
from the putrefaction with the free interchange of atmosphere between the house
and the foundation would insure the immediate removal of the cause of the
stench.

This mass of broken stone, with its
putrefying carrion below and its human
habitation above, is only an exaggerated
illustration of what exists universally over
wide ranges of country. Houses are
sometimes built on coarse gravel. Here
the atmospheric interchange is almost as
free as in the illustration given. Sometimes the gravel is finer and mixed with

sand which, imposing by friction more
resistance to the movement of the air,
limits the interchange ; but interchange
to the extent of free inhalation and
exhalation always goes on. Nothing
can prevent this from being active
when chimneys are drawing strongly,
while the house is sealed against the
outer air; when, indeed, as is so often
and so widely the case on light soils, the
whole practical ventilation of the house—
that is, its intake of air—is from the
ground under it, often flowing through
and enriched by the various familiar fumes
of ill-kept cellars.

The putrid carrion, it is true, we do
not find in such concentrated condition as
to produce an insufferable stench ; but let
us examine the case of a certain village.
It is not necessary to name it. There is
not a State in New England in which
many of its parallels may not be found,

and, indeed, there is hardly a village in the whole country built on a porous soil where corresponding conditions do not exist.

The village that I have in mind was built on a flat deposit of gravel inter-mixed with very coarse sand, lying nearly level and extending in depth about fifteen feet to the permanent level of the adjacent tidal waters. It was a considerable village throughout the first half of the century; then it began to expand into an important railroad town. It has now a large popu-lation and much wealth. It has a water supply, and "all the modern improve-ments"—all except sewers. Its disposal of household waste of all kinds is not *upon* the soil, which would be obviously indecent, but *into* the soil, which has the supposed advantage of hidden indecency.

The result must inevitably be a diffusion throughout the whole underlying groundwork of the village of putrefying kitchen grease, and fecal matter and laundry slops, which can not fail to produce in the whole atmosphere of the gravelly earth a condition of marked contamination. Even in the milder season, however free the interchange between the air in the ground and the air over it, the air of so much of the ground as lies under houses can not be by any means ideally perfect. When the interchange between the outer air and the ground is cut off by frost, and when cellars and wells form almost the only means of communication, then the condition is only infinitely worse.

This description may seem at first reading too sensational, and dwellers on light soils will point with satisfaction to the relatively low death-rate that their communities furnish as contrasted with that of dwellers on damp clay soils, where this atmospheric interchange is practically inoperative. This is no fair response. The death-rate is comparatively low under these circumstances, not because of, but in spite of, the almost universal breathing of the products of putrefaction as exhaled by the soil into the house. Could this element be withdrawn, it can not be questioned that, on the lighter soil, the death-rate, and in larger degree the sick-rate, would show a much greater contrast.

The practical question now arises, how to meet this difficulty? If proper sewers were once provided, an absolute suppression of all vaults and cesspools would suffice to secure the early purification of the ground, for the bacteria of decomposition —those universal scavengers—would soon make away with the existing accumulation. How far their action may modify the present ill effects of the constantly renewed underground filth we have as yet no means of knowing. If we are wise we

shall take the benefit of the doubt and cut
off the supply of foul material.

Sooner or later, we shall secure, by
sewerage and a compulsory use of the
sewers, the complete purification of the
subsoil. In the meantime, the individual
householder who has an anxious thought
as to the condition of his individual
house, and who is now living subject to
the influences of an evil due to his neighbors' many cesspools more than to his
own single one, should seek some means
to protect himself against enemies which
his neighbors are willing to disregard. He
will find his best protection in isolating
his house in the most effective way from
the ground in which it is founded. There
is a common belief that stone walls laid
in mortar, and cellar floors covered with
a few inches of concrete, effect such isola-
tion. This is not the fact. Concrete
floors and granite walls are as sponge to

the penetration under slight pressure of atmospheric currents. To what degree walls and concrete floors filter out the impurities of the air passing through them we do not know. Not knowing, we will not trust.

One of the safest materials for a cellar bottom, and for the exterior packing of foundation walls, is a clean, smooth, compact clay, one which may be beaten into a close mass, and which has a sufficient affinity for moisture always to maintain its retentive condition. When used in the damp atmosphere of a cellar or about a foundation, it seems to constitute a good barrier to the passage of impure air. In the cellar it may, of course, be covered with concrete for cleanliness and for good appearance; but six inches of clay well rammed while wet will impede the movement of air to a degree with which ordinary cellar concrete can furnish no parallel. Where clay is not available, a good smearing of asphalt over the outside of the founda- tion-wall, and a layer of asphalt between two thicknesses of concrete for the cellar- bottom, will afford a complete though more costly protection. Asphalt used in substantially the same way, especially if in connection with a solid course of slate or North River bluestone, in the founda- tion above the ground level, will prevent the soaking up into the structure of the moisture of a heavy soil.

The matters above touched upon are seldom discussed in works on house- drainage, except so far as the mere removal of surplus soil moisture is con- cerned, but their importance is not likely to be overestimated.

There may be good grounds for the opinion of those who think that many of the minor ailments to which the race is subject, and some of its more serious ailments as well, are due, not to the influence of an excess of filth in any form, but to the influence of an excess of moisture acting often on a little filth, or on a little organic waste which would not be classed as filth at all. Such ailments prevail more especially in houses in which mold is prevalent, which on being closed soon acquire a musty smell, and in which stuffiness is a natural condition—houses where a general and all-pervading slight dampness is to be detected. This dampness may belong to the structure rather than to the climate; for there are dry houses at the sea-side and damp houses on the mountains.

The soil has an influence over the interior climate of the house, which is even stronger than external atmospheric conditions. Positive knowledge does not carry us very far in this direction, but the experience and observation of the world, especially where intermittent fevers and neuralgia prevail and where an ailing condition and low tone are the rule, have indicated very clearly that the wisest course for every man who would make his home perfectly healthy would be to separate it as completely as possible from all interchange of air or moisture with the ground on which it is built.

CHAPTER VI.

FOUL DRAINAGE.

ANOTHER and even more important branch of house-drainage has come into general use within a comparatively short time. This is now attracting quite all the attention that is its due. Knowledge concerning it is advancing steadily, and on the whole satisfactorily. Mistakes have been made during the past dozen years even by the best of those who have had to do with it. Such mistakes have from time to time become recognized, and they have been remedied, until we are now approaching something like a fair understanding of the fundamental requirements of house-drainage. Perhaps it would be too much to say that the practice of the art keeps any thing like even pace with the knowledge as to its principles.

Neither the common usage of the best plumbers nor the average requirements of the boards of health of cities show any very considerable improvement over what was done in the better work of some years ago, save in better workmanship.

Leaky joints in iron pipe, though still by no means uncommon, are less frequently found since attention has been given to testing joints under pressure. In the best work, the thorough ventilation of soil-pipes, furnishing an inlet as well as an outlet for the movement of air, is now generally adopted. Another step in advance is marked by the abandonment or the much better construction of drains laid under cellar-bottoms.

The greatest step of all—the step which insured wide public benefit—was taken when municipal boards of health became so generally, so almost universally, interested in the subject of plumbing regulations. These bodies have nearly every where established an effective control over all new work done, and often over the amendment of old work. The main point being gained, that all such work is to be executed according to rules and under such inspection as will secure the observance of the rules, it is only a question of time when the rules themselves shall be perfected.

As they stand, these plumbing regulations permit some things which they will hereafter prohibit, and they require some things which they will hereafter, perhaps, not permit. In the latter category is the back ventilation of traps, and in the former the use of " pan " water-closets, of fresh-air inlets at the level of the sidewalk, and of bends, cowls, and caps at the top of the soil-pipe.

However, in spite of all their imperfections, the establishment of such regulations, and the rigorous enforcement of their requirements under actual inspection, have marked the greatest progress that has been made for a long time past. It is to be remembered, in criticising these regulations, that they are necessarily made suitable for universal application. They are a very inadequate guide for the arrangement of the plumbing work of a large and elaborate house ; but they do constitute an invaluable guide and safeguard for work of a cheaper sort. The poor tenant, who was formerly at the mercy of his landlord, is now protected by a system which must inevitably prevent the repetition of the infamous work of the cheap plumber of a few years ago.

NOTE.—Fortunately, what is here said as to the common usage of the best plumbers and the requirements of boards of health, is no longer true. Plumbers—of the better class—have improved their methods and their work very greatly, and boards of health, so far as their need to regulate their requirements to the poorer classes will permit, are exerting a most beneficent influence over the house drainage of their towns.

Board-of-health rules and regulations have been improving, year by year, and while they are still often far short of perfection, there is good reason to be satisfied with the progress that is being made.

"Pan" water-closets are quite generally prohibited, but the back ventilation of traps and the sidewalk air inlets still hold their sway.

CHAPTER VII.

I T is no part of the purpose with which this book is written to discuss, even in a general way, the different methods and processes of house-drainage, nor the various theories and opinions by which different writers on the subject are influenced. It will be assumed that the reader will be satisfied to find here only the writer's own opinion, and a statement of the grounds on which that opinion is based. I shall therefore confine myself to saying what I advise doing, with the reasons therefor.

I advise, above and before all, that in every house, large or small, the *amount* of plumbing work be reduced to the lowest convenient limit ; that there be not two sinks or water-closets or bath-tubs where one will suffice for reasonable convenience ; that under no circumstances shall there be a wash-basin or any other opening into any channel which is connected with the drainage system, in a sleeping-room, nor ordinarily in a closet opening into a sleeping-room. I should confine all plumbing fixtures on bed-room floors to bath-rooms.

I should give each bath-room exterior ventilation, but I should never locate its pipes against an outer wall unless I could give adequate protection against frost, for the liability to danger from the freezing of waste-pipes, traps, etc., is greater than the liability to danger from an interior location—if the fixtures are all of the best sort, and if the room itself is sufficiently ventilated.

I should always, so far as possible,

place the bath-rooms so nearly over each
other on different floors, that they could
all be connected by short waste-pipes
with one vertical soil-pipe, or so that the
soil-pipe could reach them with short off-
sets.   If bath-rooms or water-closets were
required on all floors, or on any floor, in
different parts of the house, I should
serve each set with its own vertical soil-
pipe, avoiding any considerable horizontal run, such as is at times resorted to in
connecting fixtures at different points on
different floors.

I should try, so far as possible, to have
every part of the plumbing work fully
exposed to sight.   It is occasionally necessary to run a soil-pipe or other waste-
pipe in a position where it ought to be
concealed ; but I should, when I could,
avoid such situations, and when possible
I should resort to some frank decoration
of the pipe rather than to its concealment
behind a casing.

Wherever pipes pass through floors in going from one story to another, I should make an absolutely tight blocking of the channel. As generally arranged, the soil-pipe and other pipes run through bungling openings in the floor concealed behind carpentry of one sort or another, and the pipes themselves are boxed in so that the whole system constitutes a free run-way for vermin, and a free channel for the diffusion from cellar to garret, and between floors and behind partitions, of whatever foul air an ill-kept cellar and closet-fixtures may produce.

The diffusion throughout steam-heated and ill-ventilated rooms of the floating results of hidden decomposition is apparent to a fresh nostril in many a "first-class house." There is no minor item connected with house-drainage that is productive of such an obvious improvement in the atmosphere of the rooms as the shutting off of this means of intercommunication.

I should use only extra-heavy soil-pipe, or pipe at least with extra-strong hubs, so that the lead calking can be driven so tightly home as to make leakage under any pressure absolutely impossible.

I should try to avoid the placing of plumbing fixtures of any sort in the cellar of a house, unless they could so be arranged as to deliver into a soil-pipe or drain not concealed under the floor. In exceptional cases, where an underground drain is necessary, I should not follow the regulations and lay a mason-work trench with a movable cover, so that access to the pipe could be gained at pleasure. I should have the pipe laid in an open trench, and so thoroughly calked that under a pressure equal to the height of one story not a drop should escape at any joint ; and then, a safe conduit being secured, I should inclose it in a concreting of the best cement, embracing it so completely and so securely that if the iron should rust out and be washed away, the cement itself would constitute a safe channel.

I should make it a chief aim to secure for all needed fixtures the greatest simplicity, and for all waste-pipes the greatest absence of complication. I should use sinks without grease-traps, bath-tubs without inaccessible overflows, wash-basins free as far as possible from fouling places, and water-closets without valves, connecting rods, or machinery.

This suggestion is a radical one, and it will fail of acceptance in many most respectable quarters. There can be, however, no question as to the propriety of expressing one's firm convictions in the most distinct way. What I am endeavoring to convey is not the well-known average opinion of engineers and sanitarians—only my own opinion. This may be entirely wrong : but it is the outgrowth of the best thought that I have been able to give the subject, and it must be conceded that no harm will result to the health of the people if it is followed out in practice.

NOTE.—In connection with the instructions given in this chapter, attention is called to the note to Chapter XIII, concerning the Durham system of piping.

It is still my opinion that grease-traps, of the kind generally used, are a nuisance, and that their use should not be permitted. There is another sort of grease-trap now available, however, which obviates the deposits which it is the purpose of the old grease-trap to cause. This is the " Siphon Eduction " trap, made by Flushtank Company, Richmond, Ind. Grease and other floating matters are retained temporarily, and at regular intervals an automatic flushtank, which discharges a copious stream through a flushing rim, washes everything clean.

CHAPTER VIII.

THE SEWER-GAS QUESTION.

THE main purpose of house-drainage, as we now understand it, is to remove all such wastes of domestic life as are suited for transportation in running water with the greatest completeness and with the greatest attainable safety. To secure this object, the drainage system must be so constructed as to carry away, completely and immediately, every thing that may be delivered into it; to be constantly and generally well ventilated; to be frequently and thoroughly flushed; and to have each of its openings into the house guarded by a secure and reliable obstacle to the movement of air from the interior of the drain or pipe into the room.

It is no longer a question of "sewer-gas." Wherever the offensive exhalations designated by this term exist, wherever the effluvium of putrid waste may be detected, there is inevitably defective arrangement, or defective workmanship, or both.

It is no longer to be considered the best policy to shut off sewer-gas from the house by confining it to the sewer. The true course should be to seek the seat of the evil and to remove its cause.

The foul air in a defective sewer or in a defective house-drain—and it more often originates in the latter—is invariably the result of the accumulation and retention of filth—its retention for a long enough time to allow it to enter into putrid decomposition. There is but one proper way to cure it ; that is, to prevent the accumulation. Such removal is to be secured only by thorough flushing, either by a copious stream accompanying the discharge, or by frequent periodic washings sufficient to sweep all deposits away. No flushing will prevent some sliming of the pipes, but good ventilation will take care of this.

All drains, soil-pipes, and waste-pipes should be absolutely tight, not only against the leakage of liquid, but against the leakage of air; they should be so reached, in every part, by a flushing stream of one sort or another, that deposit and accumulation will be impossible; they should be as thoroughly ventilated in every part as the safety of the water-seal will permit. The exterior drain, and ultimately the sewer into which it delivers, should have the same general characteristics, it being understood that the freest possible ventilation is to be given to both sewer and house-drain, by the admission of air from with-out and the delivery of air to the open sky, without the possibility of its entering the house at any point, in any manner, or at any time.

All fixtures should be so trapped that the exclusion of the air of the drain should be assured, but at the same time in such a manner that at each use of every fixture all the filth that it deliv-ers shall be carried completely away, the trap being immediately refilled with fresh water.

Such are the leading sanitary require-ments of house-drainage. These being secured, it is a matter of little sanitary consequence whether the fixtures them-selves are cheap or costly, simple or elaborate, ornamented or plain. As, however, these appliances are devoted to the meaner uses of the household, good taste would indicate that their most appropriate "elegance" is to be secured by making them and their belongings as simple as possible, and as inexpensive as the securing of the best results will allow. They should be conspicuous, if at all, by their purity and cleanliness.

Having thus set forth the general principles that should govern the construction of the drainage work of houses of all classes, we may next consider its details.

CHAPTER IX.

A CHIEF obstacle to the simplification of the plumbing works of a house arises from the mistaken commercial instinct of the plumber. It looks at first blush as though any thing tending to cut down the amount of their work must be injurious.

So far as the present force of good plumbers is concerned, it seems to me that their business interest lies entirely in the other direction. New houses are being built with great rapidity and old houses are having their plumbing work rearranged more and more thoroughly every year. Certainly the work is increasing much faster than is the number of plumbers qualified to do it properly.

If the present profusion of plumbing appliances is adhered to, the only possible way in which construction and repair can be done will be by increasing the number of plumbers, faster, much faster, than material for the making of *good* plumbers will present itself.

Of course, the number of men engaged in this handicraft must increase constantly, but the more the amount of work in each house can be reduced the better and longer hold will the present force of good master and journeymen plumbers and the well-trained future addition to that force be able to do all of the good work offering within their reach. They are suffering now very greatly from competition with inferior and unprincipled men. The more rapidly the amount of work to

be done increases, the more will this competition work to their disadvantage.

If this suggestion is sound, the best
course for the better plumbers to adopt
in the interest of their own business will
be to approve of and to recommend such a
reduction of the amount of work in each
case as will enable them to manage the
growing number of cases.

It is appropriate in this connection to
offer a distinct recognition of the fact that
practically, so far as the interests of the
whole people are concerned, the plumbing
fraternity are by far the most influential of
all house sanitarians. Engineers, physi-
cians and health officers accomplish much
by their influence with individuals, and by
the exercise of their professional and
official functions ; but they reach after all
only a limited section of the community.
The plumber on the contrary makes his
influence felt on every hand. Where an

engineer or a sanitarian has to do with the drainage of one house, a plumber in good practice has practically the absolute control of a hundred houses. Ninety-nine men out of one hundred would receive the suggestion of a theoretical sanitarian in a very gingerly way, while they would accept without question the dictum of a practical plumber.

Plumbers of the better class are fast coming to recognize the fact that the prosperity which this popular confidence brings to them, carries with it a serious public responsibility — a responsibility which, as a rule, they are endeavoring to meet in a spirit that is at least rare with men of other crafts. It is not universal with them.

Another responsibility falls upon those who undertake to instruct architects, house-owners, and plumbers themselves as to the proper management of house-drainage, i. e., the responsibility that attends an interference with plumbers' work. It is not only important to prescribe what should be done, but it is important to do this in such a spirit, and with such clearness, as to carry persuasion and conviction to the minds, and to engage the willing interest, of these ubiquitous guardians of the health of the household. It is mainly because of this responsibility that I am anxious to assure those whose interests lie in the construction of house-drainage works that the simplification of drainage systems which I so earnestly advise is in no respect inimical to the best business interests of the trade.

CHAPTER X.

IN arranging the details of house-drainage the main line is always first to be
considered. It begins at the sewer, or
flush-tank, or—in barbarous instances—
at the cess-pool ; passes through the
house by such a course as may be indicated by a judicious compromise between
directness and convenience, past the location of the highest fixture that is to discharge into it, and then passes out through
the roof for free ventilation.

The question of a main trap between
the house and a public sewer has been
much discussed, and is still determined
by no rule. There should always be such

a trap between the house and a flush-
tank or a cess-pool. I am inclined to the
belief that there should not be such a trap
in the case of discharge into a sewer,
unless it be especially foul. If it is only
a great cess-pool, holding the accumulated deposits of a street or larger district,
or if its interior atmosphere is at all comparable in offensiveness with that of a
cess-pool, then a trap will be desirable ;
but if it has such an atmosphere as will
admit of the entrance of workmen, and if
its contents are carried forward in its current with reasonable completeness, I in-
cline to the opinion that, even if no other
house connected with it aids in its ventilation, it will be better that the single
house under consideration should be connected without a trap.

I have reached this conclusion slowly
and in opposition to the opinion of many
of the best engineers. The objections

ordinarily raised against the practice are that by it "the sewer-gas is laid on" to the house ; that contagious diseases existing in other houses connected with the sewer will communicate their infection directly to any house not so cut off ; and that, as a matter of common policy, one man alone should not ventilate a sewer that is used without ventilation by neighbors.

There are two arguments against this, and they seem to be controlling ones. (*a.*) The purpose to be secured is the greatest practicable purity of the drains and pipes of the particular house, and, while it is true that a trap will shut off the air of the sewer, it is also true that the trap itself, unless the course of the drain is very steep and its flushing very copious, may not only form a seat of decomposing filth, but will so set back the flow as to cause a deposit of foul material for some dis-

tance along the drain on the house side
of the trap.

If the sewer is not extremely offensive—more offensive than a critical investigation made a few years ago showed
most sewers in New York city to be—
there will be less stench coming from
a current of air flowing from the sewer
without a trap than will be devel-
oped in the house-drain itself with a trap.
The absence of the trap will secure a
pretty constant and effective current of
air from the sewer through to the top of
the soil-pipe. With the trap, a sufficient current can be established by the
use of a well-placed fresh-air inlet ; but
the immediate seat of decomposition in
and behind the trap will continue active.
(*b.*) All the cry about sewer-gas being
"laid on," and about the intercommunication of diseases from one house to
another by means of the sewer, is the out-

growth of a condition that is now hardly tolerated, and that certainly is not contemplated in this paper. In the older work, there was either no ventilation whatever to the drainage system of the house or it was very inefficient. The water used, though perhaps not less in amount then than now, was not so used as to secure a good flushing effect, while the stability of traps was then little thought of.

Pressure of any sort being brought to bear on the atmosphere of the sewer, foul air escaped into house-drains and found no other means of relief than by forcing traps or by working its way out at defective joints. Under such circumstances, the argument in favor of the trap was a strong one. Now, house-drain and soil-pipe are tight, ventilation is very free and complete, the effect of a pressure on the air of the sewer is not to be feared, traps are reliable, and, in the best work, joints are absolutely tight.

Under such conditions the safeguard supposed to be furnished by the exterior trap is not needed—assuming always that the sewer is a reasonably clean one. Its condition will always be improved by the ventilation furnished by the untrapped drain.

CHAPTER XI.

IN the case of country houses, not discharging into sewers, the trap is a
necessity. Wherever a trap is used, there
must be on the house side of it an inlet for
fresh air. There can be no real ventilation of the drainage system if it is open
only at its top. A bottle can not be ventilated by removing its cork, nor will a
chimney draw if it has no opening at the
bottom. A copious inlet for fresh air,
working in conjunction with a wide opening at the top of the soil-pipe, will insure
a free movement throughout the whole
system that will accomplish an adequate
ventilation, not only of the main channel

itself, but, by the diffusion of gases, of *short* branches connecting fixtures with it.

Most of the directions given in sanitary journals and books for the arrangement of fresh-air inlets, especially in cities, seem to have been made without due regard to their liability to become obstructed by rubbish, and especially to become entirely closed by accumulations of snow. Many such inlets in New York, at the edge of the pavement or at the face of the curb, are sometimes blocked for days together in bad winter weather. Becoming obstructed from any cause, their efficiency stops, and for the time being the security that they should afford is withdrawn.

There is really no good reason for placing this opening at a distance from the house. I have never known annoyance resulting from the inlet pipe being brought out at the face of the foundation wall, preferably, of course, not too near to windows and doors. With well-flushed pipes, as already said, the constant though often slow movement of air through them so reduces the offensiveness, which a few years since was thought to be inevitable, that, although there might be a slight outward puff when closets or baths are discharged, no annoyance results.

CHAPTER XII.

WHETHER the soil-pipe passes through or under the foundation of the house, unless the wall be old enough for all danger of settlement to have passed, it should be carried through an arched opening to prevent its disturbance if settlement does occur. In any case, the iron pipe should be continued for nearly or quite a full length (five feet) outside of the foundation wall. It may be continued further with advantage. Although thus laid in the ground and used as a drain, iron pipe is not, like earthenware pipe, imperishable; still the greater certainty of tightness, and correct grading, if due only to the better class of workmen by whom it is done, is a strong argument in its favor. After reaching solid ground that has not been disturbed in excavating for the foundation, *a carefully laid and rigidly inspected* earthenware drain is to be preferred.

After the drain passes inside of the foundation wall it is better, where it is not necessary to connect with fixtures in the cellar, that it should be carried in full sight, along the face of the cellar wall or suspended from the floor-beams, to the point where it is to turn up as a vertical soil-pipe. This is advisable because here as much as anywhere else in the house, it is important to be able to inspect the joints, and to know always the condition of the work.

If, however, it should be necessary to make connection with a water-closet or other fixture in the cellar, it is better that the main channel should run under the floor to or near the location of such fixture, in order that all or nearly all of its length may constitute a part of the main line, thoroughly flushed and thoroughly ventilated, like the rest of the system.

If there are several vertical soil-pipes, it will suffice, of course, if one of them is carried down for the cellar connection, and the others can be carried together above ground and connected with the main line before leaving the house. A branch only ten or twelve feet long, running to a servants' closet in the cellar, even if provided with adequate upward ventilation, is not likely to keep in nearly so good condition as it would if carrying also the discharge of closets and baths above.

Whenever it becomes necessary to lay the drain under the cellar floor, I should *not* counsel the following of the usual recommendation to lay an iron pipe in a mason-work trench, with a cover that may be removed for inspection. It should be protected as hereinbefore described.

CHAPTER XIII.

THE SOIL-PIPE.

IT is a generally accepted rule, and a good one where space suffices, to use no short turns—technically, "T branches" and "quarter bends." Two one-eighth bends, or a Y branch and a single one-eighth bend, give a more gradual and therefore better change of direction. So, in the attachment of water-closets to vertical soil-pipes, it is usual and better to make the connection with Y branches. Where space does not suffice, however, a half Y answers a sufficiently good purpose, and even a T branch (right angle) is less objectionable than it was when flushing was less copious than now. The soil-pipe throughout its whole length, horizontal as well as vertical, should be so secured with hangers and clamps or hooks and with supporting posts that it will be rigidly fixed in its position.

From the beginning of the work, every joint should be made with a view to being tested under hydraulic pressure. If the workman has this in view, the test will generally discover few leaks. As ordinarily made, especially where the whole circumference of the pipe is not easily accessible to the calking tool, a test will almost invariably disclose serious leakage.

In every case the test *should* be made, and every semblance of a leak should be calked until thoroughly tight under pressure. In making this test, the simplest way is to close all openings into the pipe with disks of india-rubber compressed between two plates of iron forced together

with a screw. Such plugs can be fastened
so tightly as to hold a head of fifty feet.
There is no special advantage, however, in
applying this force ; for if joints are to
leak at all, they will leak usually under a
head of a few inches, and always under a
head of a few feet. It is generally most
convenient to test the vertical pipes story
by story, the plugs being inserted through
the water-closet branches.

Another satisfactory test which may be
applied after all fixtures are attached, is
made with an air-pump and pressure-
gauge, such as gas-fitters use. If the
gauge stands firm even under a slight
pressure for an hour together, the work
may be accepted as tight. The principal
drawback is that, if the joints are not
tight, it is much more difficult to locate a
slight leak than when the water test is
used.

I think it may be accepted as a well-grounded rule that no prudent owner should receive and pay for his plumbing work until all of the iron waste-pipe has been tested, by one or the other of these methods, under the personal observation of the architect or his plumbing expert. There is probably no occasion to fear that work once made tight will develop leaks for many years, the tendency to rust after a time, even with tar-coated or enameled pipe, being rather to close such slight leaks as may exist.

The fear has sometimes been felt that sand-holes and slight imperfections in cast-iron soil-pipe may lead to the permanent injury of the work. Ordinarily, this is not a real danger. Where pipes have been tested before erection by being filled with water in single lengths and rejected because of slight leaks, it has been found that a few hours later such leaks have become entirely closed with rust. Doubtless a rust closure is a permanent one.

The use of cast-iron for soil-pipe and minor waste-pipes is still almost universal. They are jointed with lead, which should be of sufficient amount to fill the joint entirely full at a single pouring. This lead is then tightly driven home with a steel caulking iron, and so packed closely against the walls of the pipes.

There are two grades of soil-pipe known to the trade, " common " and " extra-heavy." If common pipe has sufficiently strong hubs to stand heavy calking, *and if the outer and inner circumferences are concentric,* there is no reason why it may not be trusted for very long service ; but it is difficult to maintain the core in a perfectly concentric position, and even in the best pipe there is generally a slight difference of thickness

between one side and another. A very
slight difference is a serious matter in
common pipe. In extra-heavy pipe, unless the eccentricity is very obvious, even
the thinner portion will be thick enough
for safety. This thicker pipe, however,
is sometimes weakened by air bubbles in
the mass. To detect these, the pipe
should be tested by sharp hammering
over its whole surface.

In ordinary work in private houses, a
diameter of four inches has been adopted
as sufficient for the soil-pipe. So far as
the mere water-way is concerned, this
diameter is ample, even when roof water
is admitted from very large houses.
Indeed, for most cases a diameter of
three inches will furnish a sufficient
water-way ; then, again, the smaller the
pipe the more thoroughly it is flushed by
the stream discharged through it.

There is, however, another consideration that is important. The siphonic action, or suction, produced upon lateral branches by the discharge of water through the main shaft, is in inverse proportion to the diameter of the pipe. The sudden discharge of a water-closet using three or four gallons of water through the three-inch soil-pipe might, under favorable circumstances, produce a considerable vacuum in the branches. The same volume flowing through a four-inch pipe would have a less effect, and through a five-inch pipe still less. Practically, where there are no fixtures higher than the fourth story, and where the admission of air from the top of the soil-pipe is very free, four inches may generally be regarded as a safe size.

NOTE.—For a long time what is known as the Durham system of soil and waste piping was too imperfect in its details to satisfy the requirements of the best work.

Mr. Durham has now, however, greatly improved the

castings used for jointing his pipes ; has substituted steel
for wrought iron ; and has perfected his system in all its
details. It is now much the best method of piping, and,
especially as it is not very much more costly, it ought to
supplant the use of cast-iron pipes in all but the cheaper
classes of plumbing.

The whole work is thoroughly galvanized, inside and
out, all the joints are screwed together with deep threads,
and, aside from the better practical result, it has a much
better appearance, which is not a small matter now that
it is the rule to leave all the piping exposed to view. As
a rule, the single pipes are long enough to reach from
floor to ceiling, and the joints and branches are out of
sight under the floor.

In the reconstruction of the plumbing of the United
States Capitol at Washington, in 1893, I used the Durham system exclusively. There was over a mile and a
half (8582 feet) of piping, ranging from 8-inch to 2-inch.
The work was necessarily very complicated, and there were
no fewer than 4742 joints. Every part of the system was
tested under an air pressure of 10 pounds per square
inch, and every joint was screwed up until it was absolutely tight ; 2008 feet of the pipe (from 3 inches to 5
inches diameter) is of brass, iron-pipe thickness, and 1422
feet of this—all that is in sight—is nickel-plated. The
remaining pipe, in the cellar, is of heavy galvanized steel.

CHAPTER XIV.

THE upward extension of soil-pipe for complete ventilation is a matter of much importance, and one that has been considerably bedeviled by invention. Experiments instituted to demonstrate the utility of different caps or ventilating cowls have not yet been carried to a complete scientific result ; but they have sufficed to establish two important points.

One is, that every ventilating cowl of whatever kind, and of whatever effectiveness during positive winds—when no cowl is needed—is invariably an obstructor of the movement of air during calms or under light winds. It is known that every deviation from the straight line obstructs the current by increasing the friction. Therefore, the cap or bend or cowl, one or another of which is almost always used, is of no real utility in a high wind, and is an absolute obstructor during light winds and calms.

The best result will always be obtained by running the soil-pipe straight up to a certain elevation above the roof—more or less according to the exposure—and leaving it entirely open at the top. To prevent intentional or accidental introduction of obstructing objects, it is a good practice to insert, and to secure, into the open mouth the ordinary spherical wire-basket that

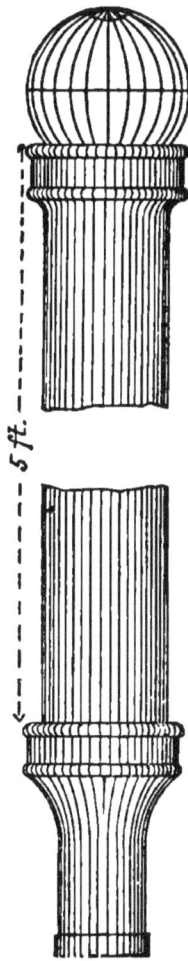

FIG. 3.—THE TOP FINISH OF A SOIL-PIPE.

is used to keep leaves from obstructing the outlets of roof gutters.

The other point is, that a universally
effective increase of the movement of air
is secured by increasing the diameter of
the pipe at its upper end. Theoretically, the lower down the enlargement
begins, and the greater it becomes at the
top, the better will be the current produced. Practically, it seems to suffice to
increase the diameter of the single
upper length of pipe. This is most conveniently done by using an " increaser,"
from four inches to six inches, just under
the roof, and to a set length of six-inch
pipe at the top.

The owner and the architect, and all
who are interested in securing good work,
should bear constantly in mind the
importance of making this main channel
for ventilation and for drainage absolutely and permanently good from bot-

tom to top. This being assured and tested, the various fixtures or plumbing appliances may be connected with its branches.

CHAPTER XV.

TRAPS, constituting one of the most essential elements of plumbing work, have for some time past occupied the careful attention of all who are interested in the improvement of house-drainage. Few who have applied their ingenuity to the subject have failed to invent and patent a " sewer-gas " trap. I took out a patent for a trap of this sort myself some years ago—probably one of the least successful of the whole list. The best of the efforts of others, thus far, have been only measurably successful. I am still using one or two of them in my own work, because they are passably good, and because

nothing else has offered that seemed better. The successful accomplishment of
the object in view offers probably the
most hopeful field to which sanitary inven-
tors can now turn their attention.

Devices intended to meet existing difficulties have not all been confined to the
form and construction of the trap itself.
Much the most widely recommended and
successfully enforced effort to meet the
difficulty has been to supply what is known
as the "back ventilation" of traps. Having known of the early failure of this
device, before it was generally recommended to the public and taken up in the
compulsory regulations of health boards,
and having carefully watched its development, I have never been able to look
upon it with favor. In some cases it
does good, but I believe that *on the
whole* it does more harm.

Not only as confirming my own view,

but as an illustration of very thorough
and careful experimental work, attention
may properly be called to an investigation
carried on for the City Board of Health
of Boston, by J. Pickering Putnam, Esq.,
an architect of that city. These investigations have been set forth quite fully in
illustrated communications to the "American Architect," which papers certainly
mark a very important step forward in
sanitary literature. The deductions to
be drawn from these investigations are
these :

While a sufficient vent-hole at the
crown of a trap will prevent its contents
from being withdrawn by siphonage (suction), insufficiency in such an opening,
resulting from whatever cause, defeats the
purpose for which it was made. Insufficiency may be due to several things. (*a.*)
The opening may originally be made too
small. (*b.*) It may, and very often does,

become reduced in size, or entirely closed
by the accumulation of foul matter thrown
into it during the use of the trap. (*c.*)
As its efficiency is due entirely to the admission of air fast enough to supply the
demand for air to fill the vacuum caused
by water flowing through some portion of
the pipe beyond the trap, it is not only a
question of having an opening large
enough to admit the air, but of having an
adequate current led freely to the opening.

As the opening is into a portion of the
drainage system that is unprotected by a
trap, it can not, of course, communicate
with the interior atmosphere of the house ;
it must be connected by a pipe either with
the open air outside of the house, or with
the air of the upper part of the soil-pipe,
above all fixtures. The ability of this pipe
to transmit air in the volume required depends on its size and on its directness.

A one-inch pipe, one foot long, for example, may admit air fast enough, while a
longer pipe of the same diameter, or a
smaller pipe of the same length, would
not do so.

One or other of the defects above indicated may very easily defeat the object,
and, in so far as the opening may be
decreased by the accumulation of waste
matters, the object, which is fully secured
while the work is new, may be permanently defeated by a condition that occurs
after a little use.   What seemed originally
to be adequate security may become un-
trustworthy in time.

Then, again, the trap to which such
back ventilation is applied depends for
its efficiency on the permanence of its
water-seal.   A water-seal which has no
other exposure to the air than it gets under ordinary circumstances, will not be so
reduced by evaporation as to lose its value

for a considerable period ; but with back ⚹
ventilation, a current of air is established
through the pipe in the immediate vicinity
of the trap, and evaporation becomes
more rapid, destroying the seal by removing the water, in a very short time. It
was an unsealing due to evaporation that
first caused me to discard the method. I
believe, most firmly, that when the system
of back ventilation, as now practiced, is
applied to all the traps of a house, the
destruction of the seal, by evaporation,
will be much more to be feared than it
would be in the same set of traps by
siphonage only, if not vented.

Traps are also frequently emptied of
their water by capillary attraction. When
a rag, a bit of string, a matting of hair,
or any other porous substance having
one end immersed in the trap, has the
other end extending over the bend and
leading into the discharge pipe, traps

having a seal of only the ordinary depth may be emptied in a short time by this action alone. In other cases, and even where the traps are considerably deeper, the capillary material, by increasing the evaporating surface, greatly increases the liability to evaporation in the presence of the current of air produced by the venting-pipe. While, therefore, this capillary action is not an infrequent source of the failure of a trap which is not ventilated, its effect is much more serious when the trap is ventilated.

Mr. Putnam's experiments were conducted in logical order. He first demonstrated that the air rushing through the trap to supply a vacuum caused by a flow in the piping beyond carries the water with it as a matter of course. Some of this water, striking against the walls of the trap, is thrown back to its original position, so that the whole vol-

ume of sealing-water is rarely removed
with a single motion, whatever the form
of the trap. However, he found that,
sooner or later, under a sufficiently continued movement of air, the water,
even in a deep trap, might be so withdrawn as to break the seal permanently.
The time required for this depends very
much upon the number of surfaces of
the wall of the trap tending to throw the
water back into it. It was found that, of
the common traps, the ordinary " pot " ⟍
or " bottle" trap offered the greatest obstacle to siphonage. It was assumed that
" the severest test for siphonage to which
a trap could possibly be subjected in
practice would be that which would be
sufficient to siphon out an eight-inch pot-
trap or a ventilated S trap constructed in
the usual manner."

The apparatus used was strong enough
to destroy in one second the seal of a

one and one-quarter inch S trap, with
a one and one-quarter inch vent-opening
at the crown, having a one and one-quarter inch smooth lead pipe, sixteen feet
long, connected with it ; and to siphon out
an *unventilated* pot-trap eight inches in
diameter, having a seal four inches deep.
It was shown by this apparatus that a reduction of diameter of the vent-pipe, or
an increase in its length, lessened the
stability of the trap. It made a marked
difference whether the pipe was straight
or was bent into a coil three feet in diameter. It would seem from the description that the vent-opening was as large,
and the vent-pipe described above as
large, as short, and as straight, as would
ordinarily be found in practice ; and it
was shown that the seal was, in nearly
every case, easily destroyed.

The experiments demonstrated that
none of the ordinary traps can withstand

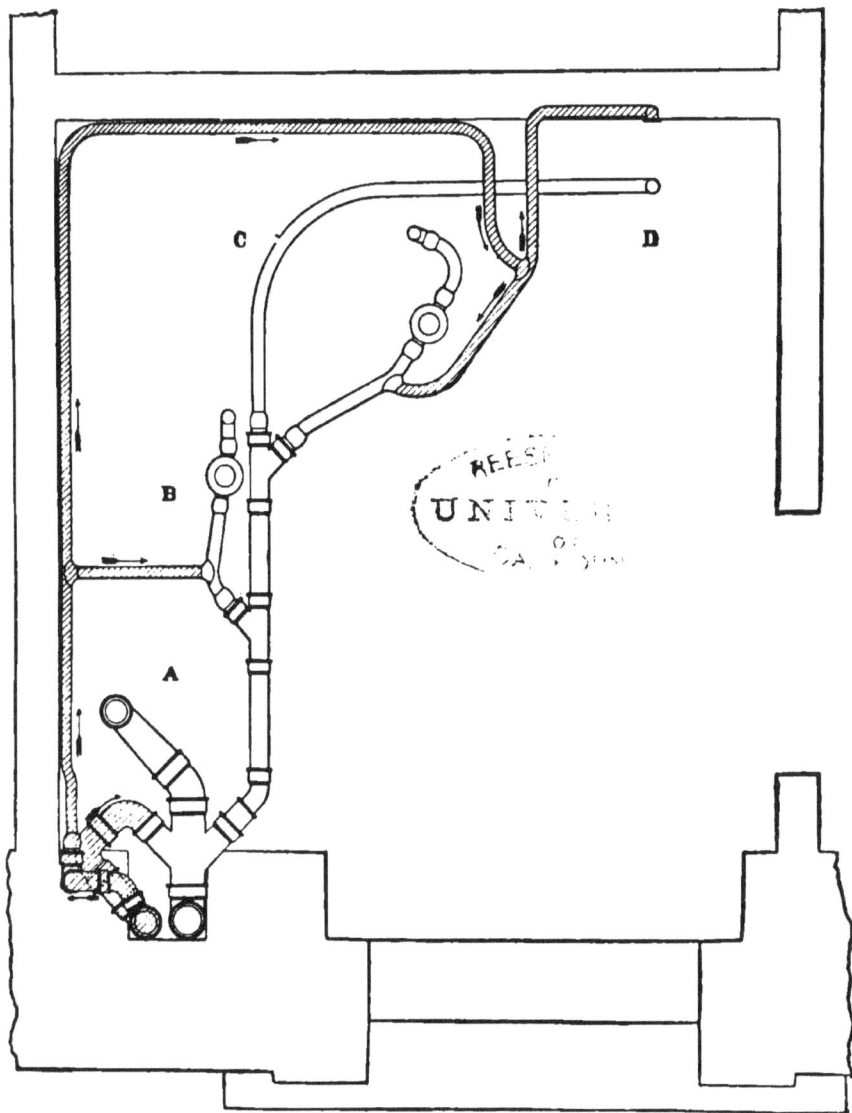

FIG. 4.—THE CROOKS AND ANGLES OF TRAP-VENT PIPES.

NOTE.—The shaded pipes are the vent pipes. The arrows show the direction of the air current as moving to prevent siphonage. A. Water-closet. B. Bath. C. Bidet. D. Wash-bowl.

a not unusual siphonic action, even with
what would be considered adequate ventilation. These experiments were repeated
in a great variety of ways with the same
general result.

If the reader will examine the tortuous
course of the various pipes shown in the
accompanying illustration (Fig. 4) of
actual work, or in any such illustration to
be found in sanitary journals, he will see
that the difference between a straight line
and a coil three feet in diameter is as nothing compared with what constantly occurs
in practice where the vent-pipes turn in
and out and up and down, and are inter-
rupted by frequent branches to such an
extent as to increase very greatly, indeed,
the difficulty of the rapid passage of air
through them.   I have seen a single
vent-pipe, having three branches on each
of two upper floors, carried down by an
irregular course with sharp turns to the

traps of a bath, a sitz-bath, a bidet, a
wash-stand, and a water-closet on the
floor below.

Aside from this serious objection, it is
not every plumber who is able to keep his
head in carrying out such complicated
work, and we frequently see a distinct
"by-pass" leading from the drain directly
into the apartment.

In tests of capillary action, the following results were obtained: Strips of hair-
felt, closely resembling the matted accumulation of short hairs which forms so
large a proportion of deposit in traps and
pipes, were used, having one end immersed
in the water of the trap and the other
hanging over the bend.   Other materials
were similarly used.   The result of the
experiments, as affecting the question of
ventilation, is thus set forth :

"To test the loss by capillary attraction on ventilated S traps, as compared

with the loss on the same where unventilated, an S trap having a seal of four and five-eighths inches was arranged as before with jute half filling the trap. With the trap attached to a waste-pipe, and connected with the drain in the ordinary manner, but unventilated, the loss by capillary attraction was as follows : * In the first five minutes, one-half inch ; in the first forty-five minutes, one inch ; in twenty-four hours, three inches ; in three days, three and one-quarter inches ; in four days, three and three-eighths inches. Thereafter no perceptible change took place. It made no perceptible difference whether the basin side of the trap was opened or closed, showing that evaporation in an unventilated trap is practically almost imperceptible.

---

* A part of this was probably due to the absorption of the water by the fibers of the jute.—G. E. W., JR.

The experiment was then repeated on
the same trap ventilated at the crown into
a cold flue, with the following result : In
one hour, one and one-eighth inch had
been removed ; in five hours, one and
seven-eighths inch ; in twenty-two hours,
two and a half inches ; in two days, three
and one-quarter inches ; in three days,
three and a half inches ; in four days, three
and three-quarters inches ; in five days,
four inches. Thus the loss continued at
the rate of about one-quarter inch a day by
evaporation, after the outer end of the
jute mass had entirely dried up. This
evaporation was nearly double *what it
would have been had it not been assisted
by the capillary attraction.* From this
we see that ventilation greatly increases
the danger arising from capillary attraction, often rendering the latter dangerous
in cases where, without ventilation, the
seal would not have been broken.

NOTE.—My experience and observation during the years that have passed since the foregoing was first published, have fully confirmed my earlier opinion that the back ventilation of traps does, on the whole, more harm than good.

In any case, there is no longer any excuse for requiring it, and the sooner boards of health cease to require it, the better it will be for all concerned—except, perhaps, the plumber.

There are now perfectly safe anti-siphon traps in the market, and in the very rare cases where these cannot be used, the McClellan mercury-seal air inlet answers every purpose.

CHAPTER XVI.

AS an incidental result of his experiments on siphonage, Mr. Putnam, by gradual stages, arrived at the invention of a trap which seems to be a practical one, and which, subjected to tests that were sufficient to break the seal of any ordinary trap even with fair back-ventilation, maintained its seal undisturbed without being ventilated. The theory followed is this : Siphonage is due to the rapid movement through the trap of air driven in by atmospheric pressure, to fill the partial vacuum formed by the withdrawal of air from the pipe beyond the trap by the inductive effect of flowing water ; the first tendency of the current thus produced is to carry the sealing-water with it.

In a perfectly smooth curved trap the removal of the water may under strong suction be complete and almost instantaneous ; in traps of irregular form, where the water in its course strikes against the wall of the trap, it is thrown back or deflected from its course ; when so thrown back a portion of the water is still carried on by the current of air, but another portion falls away from the current and resumes its position in the trap. If a sufficient number of deflecting surfaces are presented in the course of the current of air, the whole of the water, after a certain portion of the seal has been removed, is retained, and the complete unsealing of the trap can not occur.

Mr. Putnam's trap, the form of which is illustrated herewith, stands, in its normal condition, almost entirely full of water.

'Under strong siphonic action about one-
half of this water follows the air toward
the drain ; this amount being removed,
the deflecting surfaces of that portion of
the apparatus thus emptied suffice to rob

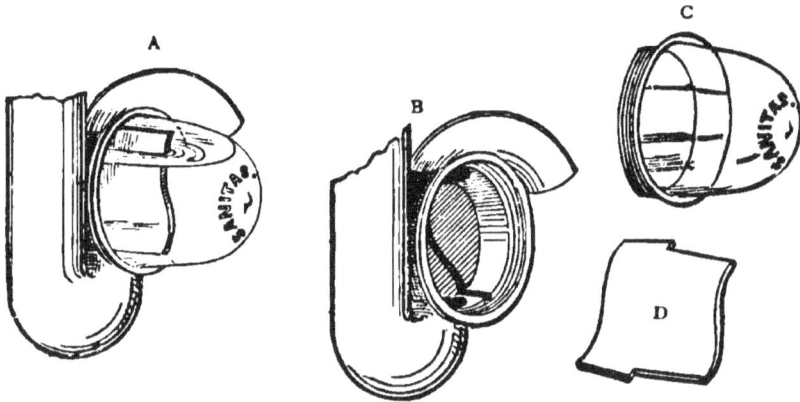

FIG. 5.—PUTNAM'S TRAP

The complete trap is shown at *a*. Its different parts are shown in the
cuts *b, c,* and *d*. The parts *c* and *d* may easily be removed for cleansing
without the aid of a plumber.

the air-current of its spray, and under no
test that has yet been applied, with an
open-topped soil-pipe, can the seal be
broken. The interior of the trap is well
exposed to view, and the arrangement
for cleaning in case of need is simple.
The trouble of an occasional unscrewing

of the glass cap to remove an obstruction would be a very small price to pay for the absolute security which Mr. Putnam seems to have achieved.*

This trap, or something like it, may probably come into universal use for wash-stands, baths, and laundry-tubs,— for urinals, also, where separate urinals are used. For water-closets it cannot take the place of the exposed trap of which the bowl constitutes one arm. For kitchen and pantry sinks I consider my own device better.

I have been using for some years past one form or other of mechanical trap, usually Bower's or Cudell's. They seemed to be the best heretofore available, but they have never been entirely satisfactory.

---

* Since the above was written, I have tested Mr. Putnam's trap, finding it effective in withstanding siphonage, and substantially self-cleansing. It seems to me the best trap that I have seen. It is entirely and permanently effective.

Continued experience with the Putnam trap has shown it to be much better than either of them.

Another excellent non-siphoning trap is the " Puro," made by The Dececo Co., of Boston. It has a smooth interior of easy curves, without angles or corners lying outside of the line of current, all parts of which are subject to the direct scouring action of the flow. The outlet side of the trap is enlarged so that it will hold enough water to afford an efficient seal after air has been drawn through the trap, and the tendency of this enlargement to retard the flow and prevent the effectual scouring of the walls is overcome by the use of a semi-circular deflector, so formed and situated as to serve a three-fold purpose : (1) It divides the flow and, while leaving enough to keep the bottom of the trap scoured by its centrifugal force, it diverts the rest obliquely toward those parts which, without it, would be unscoured because away from the flow line; (2) it prevents a retardation of the flow by contracting the waterway at the point of

FIG. 6.

greatest head; and (3) by breaking up the mass of water in the trap, it enables incoming air to reach the crown more quickly than it could if it had to struggle up through a solid body of water, thus saving to the trap a large amount of sealing water. This deflector, it will be noted from the cut, is not presented at right angles to the line of flow, but so obtusely as to offer but the slightest resistance. It cannot be regarded as an obstruction—it is rather a guide. As made, it is a component part of the trap. In fact, the Puro trap, of whatever form and whether of brass or lead, consists of a single casting, excepting, of course, the trap-screw. The absence of separable parts makes this trap very durable.

Whether compelled by local law to ventilate traps or not, I should not depend on ventilation, in the conviction that the simple S trap, as ordinarily constructed and as ordinarily ventilated, is totally unreliable.

If compelled by law to construct the prescribed " back-ventilation " I should be tempted, after its completion, to make the system inoperative by closing the main ventilation pipe at some point near its upper end.

CHAPTER XVII.

PLUMBING APPLIANCES.

CONCERNING patented apparatus, it is proper for me to explain the fact that in the following pages, among other things, I set forth somewhat in detail inventions of my own, which are patented, and by the sale of which I should profit. Such a course is naturally open to criticism, and such a position is always one of embarrassment. It is the usual course to describe the various appliances in common use, mentioning one's own only incidentally, and this would doubtless be thought by many persons to be the proper one for me to pursue.

It seems to me on reflection, however, that the only justification for the writing of this book is to communicate to the public the best advice I have to offer.

My attention has been given for many years to details of house-drainage as a matter of business, not of philanthropy. I have had occasion to study closely, and to adopt and discard, one after another, a long series of plumbing appliances— things that have come up and gone down in the rapidly improving art which twenty years ago was an extremely crude one, and in which perfection has not yet been attained in all details.

I might describe this succession of improvements, and indicate the quality, promise, and defect of each. Such information may be found, by those who desire it, very well set forth in the rather copious modern literature of the subject. The space at my disposal here would hardly

suffice for a bare cataloguing of plumbing improvements.

My own devices were in no case invented with a view to securing a valuable
patent, nor for any other purpose than to
improve my own professional practice.
The few of these devices which have ap-
proved themselves to my later judgment,
and which I am now introducing in my
work, I have patented to secure an incidental commercial advantage. I shall
therefore describe them without hesita-
tion and without further comment, treat-
ing them exactly as I treat such of the
inventions of others as I believe to be
good. I shall trust to the good sense of
the reader not to misunderstand my motive.

Special appliances for carrying out the
plumber's art in the drainage of houses
are to be numbered by hundreds. Invention has taken advantage of a growing

demand for the attainment of additional security against the invasion of drain-air, and has literally run wild. " Sewer-gas " has been made to do full duty as a cause of public alarm. The shops and the catalogues and the professional papers and books are full of an embarrassing variety of all manner of devices.

Many of these inventions are great improvements on their predecessors, but many are their predecessors under new names and with new complications. Few of them have been made with regard for what seems to be the most imperative need of the work—simplicity.

CHAPTER XVIII.

SIMPLICITY.

WE should especially seek the greatest possible simplicity, not only in detail but in general scheme. While the market offers a separate vessel for each possible separate use, the wisest course seems to be to reduce the number of vessels and to concentrate the various uses as much as may be. For example, I should, whenever possible, avoid the need for urinals, slop-sinks, and hoppers, by constructing the water-closet in such a manner as to supply all of these demands in a convenient and acceptable way, thus securing, incidentally, the most frequent change of its trapping-water and the most frequent flushing of its outlet.

The urinal is almost invariably the most odorous vessel in the house. The slop-hopper is generally a receptacle for rags and rubbish, in a dark, out-of-the-way, uninspected closet ; and the sink for drawing water is, in less degree, open to similar objection.

With a self-closing faucet for drawing water, there need be provided for the protection of the ceiling below only such simple means of outlet—like a safe-pipe opening through the ceiling of the basement or over a sink or a water-closet cistern—as will carry the slight drip that may come from an accidental leak. Ordinarily there is no serious objection to arranging to draw water through the bath-cock, if this is placed, *as it should be,* at the top of the tub.

Objections to this concentration of

uses, and to the abandonment of the provision of a separate vessel for each separate use, are confined mainly to trade journals, published in the interest of manufacturers and plumbers whose profits, it is thought, might be affected by the reduction. Their argument is that "cost is secondary to ample convenience."

While it is important to avoid unnecessary cost, the economical argument is the least of all the reasons for what is here proposed. The real and controlling argument is based on the great advantage of having the fewest possible points requiring inspection and care and to secure the most frequent possible use of every inlet into the drainage system. Reasonable convenience being always kept in view, three water-closets in an ordinary house are much better than half a dozen ; and the same principle holds throughout the whole range of plumbing appliances.

CHAPTER XIX.

STATIONARY wash-stands, where they should be used at all—in bathrooms and lavatories mainly—should, like all other fixtures of the kind, have the space under the slab fully exposed to view, so that the trap and all pipes may be seen at all times, and so that neither by accident nor by stealth may there be created the hidden untidy condition which is almost universal with the tight, unventilated, inclosed spaces once so generally used.

The basin itself, as now constructed, has a hidden overflow which it is very difficult, if not impossible, to cleanse, and it has generally either a plug and chain to close

its outlet, or a side plug operated by a
knob above the slab. Both of these are
wholly objectionable. The links of the
chain and the ring and attachments of the
plug become fouled with soapy matters,
and it is difficult to cleanse them. Practically, they are generally nasty.

To shake a filthy chain in a basin of
clear water would be a very untidy pre-
liminary to ablution. This is substantially
what we do when we let water run with
some force directly into a basin in which
a dirty chain is hanging.

The side plug *seems* to be much nicer;
it is really less nice. There is a befouled
waste-pipe leading from the outlet to the
plug, in communication with a slime-coated
overflow channel rising above the plug.
This pipe it is practically impossible to
cleanse. Its filth is constantly undergoing
decomposition. Whenever the bowl is
emptied it becomes filled with air; when

the plug is closed and the bowl is filled, this air is driven in bubbles with some violence into the bowl. Not infrequently flakes of the sliming matter come with it.

NOTE.—There has now come into quite general use a " standing overflow " for the wash-basin, which obviates these old objections.

The bowl is made to slope, not toward the center, but toward the back, at a point under the edge of the slab. The outlet is placed here, and it is plugged with a short pipe, serving also as an overflow. It is raised and lowered by a simple lever above the slab, and it may be removed for cleansing. It is a great improvement on all that preceded it.

A new waste—the " Hale," opening at the center of the bowl, after the old manner—is actuated by a lever working from below, so that the basin remains smooth and unobstructed. The overflow is through the side of the bowl near the top, but it is freed from the objection set forth in Chapter XXII by the fact that the discharge does not set back into the overflow channel, having a tendency, rather, to draw a current of air through it. It is a simple and excellent device.

In my own house I use the high " pantry-sink " cock for the supply of the wash-bowl, which favors the cleanly Mohammedan practice of washing in running water.

CHAPTER XX.

WATER-CLOSETS have, naturally, been the subject of more ingenuity, and of more argument, than any thing else connected with the subject of house-drainage.

It is hardly necessary at this late date to say any thing to the limited public which reads on such subjects about the absolute inadmissibility of the almost universal pan-closet, which was so lately the great favorite of landlords and of builders, and which, in spite of its complication and intricacy, was, owing to the great demand for it, sold more cheaply, and therefore more widely, than any other.

It is enough to say that those who care for safety in drainage works should neither adopt it in new construction nor retain it where it already exists. It is not, and it can not be made, a safe water-closet. To a greater or less degree, the objections to it hold in the case of every other closet in the market which has, anywhere in the course of its outlet, any thing of the nature of a valve or moving part.

It is not an overstatement of the universal conviction of skillful sanitarians to say that the range of unexceptionable water-closets is limited to such as have a free water-way from the bowl to the soil-pipe, depending for their trapping, and in some cases for their holding of a bowl-ful of water, on an elevation of the over-flow point. These may be classed in a general way as "hopper" closets. The simplest form of this closet is a funnel-shaped vase reaching from the floor to the seat. At the bottom it is connected with an S trap, having a depth of seal generally of from three-fourths of an inch to an inch and a half. This is a cheap and good utensil for the commoner uses. It is made of earthenware or of enameled iron, and in its best form its rear portion is nearly or quite vertical. What is known as the "short hopper," made of iron or of earthenware, has a shallow bowl, with a trap rising at its side and entirely above the floor. These are the best of the cheap closets.

Pursuing the intention already announced, to avoid any thing like a cataloguing of plumbers' supplies, and referring to what has already been said about my own inventions, I give herewith, as an illustration of the better class of closets, a vertical cross-section of the Dececo closet with its trap and discharging siphon. In this closet I have tried to overcome the objections to the mechanical or valve closets, while retaining the very great advantages of a deep bowlful of water for the reception of deposits and for the suppression of odor.

The closet has a seal about four inches deep, a depth of water of nearly seven inches, disposed in the most useful way, and a sufficient submersion of the front part of the bowl. While it is possible under strong siphonage to reduce the depth of its water considerably, it is not possible, under any conditions that can occur in practice, to break its seal, the rising limb being sufficiently large to give an adequate passage to a continuous stream of air without removing the water to such a point as to unseal the trap. It has the further advantage that its seal is in full view and is always under control. When it *seems* to be right it *is* right.

The peculiar operation on which it

depends for its discharge is due to the
use of an outlet weir below the floor,
which is the invention of Mr. Rogers

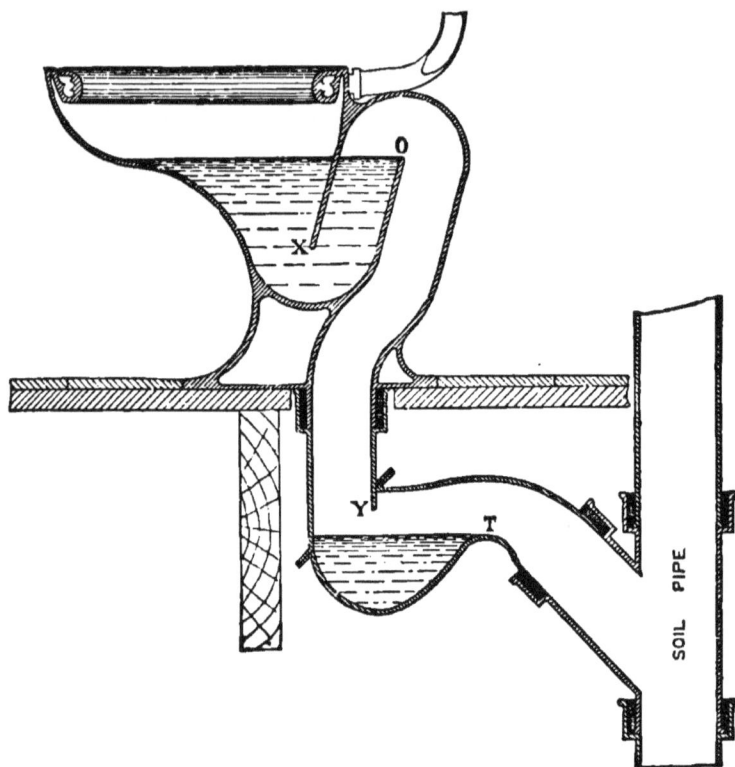

FIG. 7.—THE DECECO WATER-CLOSET.

Field, an English engineer.   It is, in fact,
a modified Field's flush tank.

The outer or discharging limb of the
syphon reaches down into the weir-

chamber. The depth of seal is the distance from the surface of the water in the
bowl to the top of the intake X, and this
is regulated by the height of the overflow
point O. The closet is supplied with
water through an ordinary flushing-rim,
connected with a service-box or cistern
overhead. The cistern is operated by a
pendent pull. When the pull is drawn
down, a copious supply of water flows
into all parts of the bowl through the
flushing-rim, washing it completely and
raising the level of its water rapidly.
The surplus overflows at O faster than it
can be discharged over the weir-top T,
without rising so high as to close the
opening at Y. This closure shuts off the
air in the siphon from the air in the soil-
pipe, with which it is ordinarily in communication. The water flowing through
the long limb of the siphon, in an irregular stream, carries the air with it, and

there is soon established a strong siphon action, which continues until the water in the bowl, into which a strong stream continues to flow, descends below the top of the intake X. Then air is admitted at this point, and the flow through the siphon is checked. The discharge at T continuing, the water in the weir-chamber soon falls sufficiently to allow air to enter at Y and empty the siphon. The water in that part of the siphon between X and O falls back and establishes an immediate hydraulic seal at the intake. The service-box is so arranged that after the main supply is stopped a small stream continues to be discharged into the bowl until it is filled to the height of the overflow point.

It was evident from long and successful experience with Field's flush-tank, that the principle on which this closet is constructed is a perfectly correct one.

It has undergone few changes since its original construction three years ago, and the several hundred closets now in use are invariably satisfactory, so far as reported.

The first ones made, five in number, were set up in the White House after the removal of the President in 1881. They are still perfectly successful in their working. After ample trial for servants' use, and for more than a year in a girls' boarding-school, an experimental one is now being tested with the rough use and rough handling of the operatives of a spinning-mill. It has, during the seven weeks since it was put in, given complete satisfaction. So far as I can see, this closet accomplishes perfectly every purpose for which a water-closet, slop-hopper, or urinal is required.

In practice, it uses at each operation over two and a half gallons of

water, which gives a thorough flushing to
the soil-pipe and drain, while it has the
great advantage of sending a good part
of its water into the soil-pipe in advance of the foul matters, lubricaing their
passage through the whole drainage system. Although this considerable volume
of water is essential to its complete
efficiency, the closet may be emptied by
pouring into it suddenly less than two
quarts of water. A large pail of slops
thrown in as rapidly as possible fails to
overflow it, and barrels of water might be
poured through it in succession as fast as
the three-inch outlet can discharge it.

The setting of water-closets in the best
manner is most easily secured when hopper or other plain closets are used ; that
is, closets with no machinery under the
seat. By the best manner, I mean such
setting as requires the minimum of carpentry, preferably nothing whatever but

a single well-finished hard-wood plank with a hole through it, resting on cleats at the sides and hinged to be turned back out of the way. It is better that there should not even be a cover to the hole. The entire closet, inside and out, should be as thoroughly exposed to view, to ventilation, and to perfect cleansing as possible.

If the floor and back and side walls be covered with glazed tiles—preferably white—so much the better ; but a cheap and satisfactory setting is secured by a slate flooring with hard-wood finish around the sides. Even oil-cloth on the floor, and the ordinary base-board and plaster at the sides and back, answer a very good purpose ; the great thing is to have a perfect exposure to sight and to air.

The costly housing-in of the closet by a close seat and cover and a close riser in

front may serve a very good purpose as
an ornamental piece of cabinet-work, but
it too often covers a condition of
things that no fastidious housekeeper
would knowingly tolerate. Sloppage,
leakage, and the tainted air rising through
the irregular holes left for the soil-pipe,
unite to make this space untidy and in
every way objectionable. Some sort of
housing-in is necessary with closets which
have machinery about them, but the
whole class of hopper closets may be
entirely free from any thing or any condition to make such concealment desirable.

It is possible that my decided prefer-
ence for the Dececo is only an incident
of paternity—but it is a decided prefer-
ence nevertheless.

NOTE.—The foregoing remarks on water-closets are
left as they were written in 1885. So far as general
principles are concerned, they are as true now as they

were believed to be then, but the general practice has changed greatly.

I am disposed to take to myself some measure of credit for this change. It has been largely due to the modern method of setting the closet, which was first applied in work that I arranged for the late E. F. Bowditch of Framingham, Mass., early in 1879. Before that time, the water-closet was concealed by carpentry, in the supposed interests of decency, and for the protection of the mechanism by which the water-supply was furnished.

The closet at Mr. Bowditch's was a simple "long hopper," standing like a white vase on a flooring of white tiles. The walls were tiled to the height of the seat, which was a single ash board, twenty inches wide, resting on cleats at its ends, standing free of the tiling at the back by some inches, hinged to turn up against the back wall, and having a hole without a cover. The previous practice having been to conceal, in the most shame-faced manner, everything connected with a water-closet, the practice was here inaugurated of exposing everything in the frankest way—exposing it to perfect ventilation and inspection, and incidentally shutting off the foul-air way, by which cellar air and the exhalations of foul sloppage about enclosed closets, sinks, basins and baths, had been freely distributed throughout the house.

This arrangement necessitated the use of a pendent-chain pull to discharge the supply-tank above, and fortunately it made the old system of valve supply impossible.

In fact, it inaugurated the practice, now almost universal, of exposing all fixtures as freely as possible—even bath-tubs now standing on their own feet, with a full sweep of air about them.

An absurd tendency is still retained, in a certain class of minds, to use a flap cover over the seat of the closet; but this relic of immodesty will pass in time, and the bowl will then in all houses, as now in the best, have the benefit of inspection and freer ventilation.

The " pan " closet has given place to the " washout " for cheaper work, and this is a vast improvement. The Dececo has many imitators, and it may fairly be said that house-drainage has undergone a complete and most beneficent revolution within the past fifteen years.

CHAPTER XXI.

SINKS.

KITCHEN and pantry sinks are used for the discharge of matters which in their original condition are not offensive. They are, therefore, in the popular estimation, of much less serious consequence to the sanitary condition of the house than are water-closets. This temporarily different condition, however, of the matters which they receive, very soon gives place to a similar condition of the matters which they have discharged.

After a little retention, putrefaction sets in, and the refuse food of the sink becomes as offensive and objectionable as does the digested food of the water-closet.

In the one case as in the other, it is very important to secure a complete removal of all foul matters well beyond the house before putrefaction. The liability to detention and deposit is much greater in case of the sink than in the case of the closet, for the reason that, with much less flushing, there is discharged through its waste-pipe a considerable amount of heated and temporarily liquefied grease.

This grease passes the strainer of the sink and is unnoticed, but, as it cools along its course, it attaches itself to the sides of the pipe in constantly increasing accumulations, until the channel is often nearly or quite obstructed. It is by no means pure grease that is thus attached. In its congelation there are involved particles of highly putrescible matters, and the ordinary kitchen-sink waste-pipe is the seat of a constant decomposition—mostly **beyond the trap, and for this reason not** especially noticeable—but especially foul nevertheless.

Not to get rid of the putrefaction, but to prevent the obstruction of the pipe, there have been invented various forms of grease-trap, having for their purpose the hardening of the grease under conditions which will allow it to be removed. These grease-traps would answer a better purpose than they do if we could depend on their being regularly attended to ; but so long as water will flow from the sink, servants will give themselves but little trouble about such accumulations. No accumulation for longer than a single day should be tolerated.

I have employed a device that has now been in considerable use for several years, which seems to meet the requirements of the case quite completely. There is built beneath the sink, and in connection with it, a " flush-pot " large enough to hold several gallons of water.   Its top is cov-

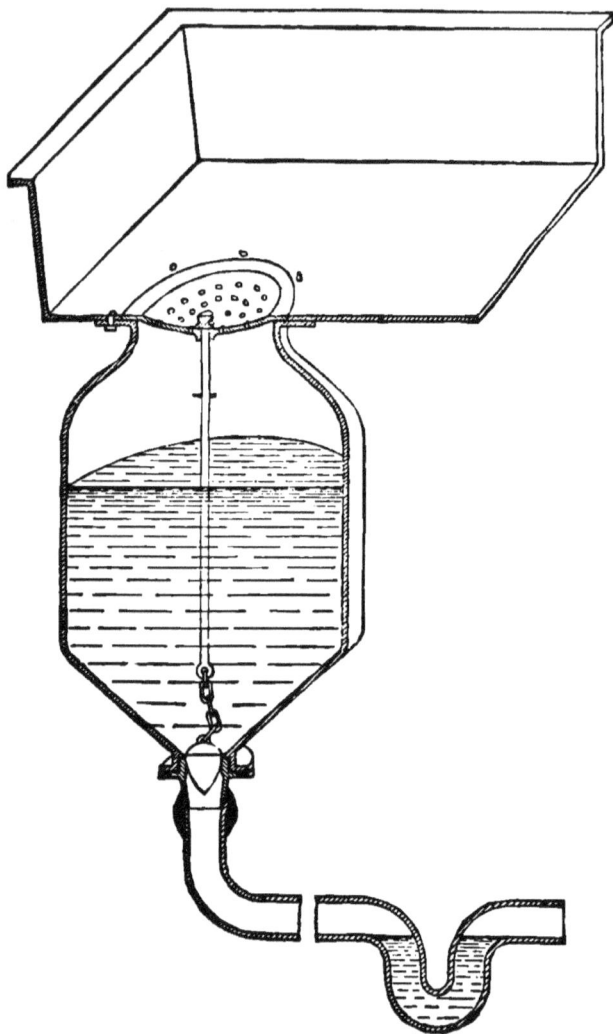

FIG. 8.—THE DECECO FLUSH-POT FOR SINKS.

ered by a strainer, about eight inches in

diameter, and pierced with large holes.
This constitutes a portion of the floor of
the sink. The outlet of the flush-pot is
closed with a plug, like a wash-basin plug,
which is attached to a spindle rising
through the strainer. The outlet is connected with the drain by a small pipe,
having a common trap, which is useful
only during the short periods when the
plug is withdrawn.

Ordinarily, the outlet stands closed.
Water thrown into the sink flows through
the strainer, leaving all coarser substances
to be brushed up and *burned in the range.**
Little by little, the flush-pot becomes
filled, and during this slow process most
of the grease becomes congealed. When
it is nearly full, the water can be seen,
even before it reaches the strainer. Then

---

*This simple cremation of the worst elements of house-garb-
age costs no money and little trouble. It solves one of the
difficult domestic problems.

the spindle and plug are raised and held
up until the gurgling of air through the
trap indicates that the pot is empty. Then
the outlet is closed and the filling begins
again. The strainer and spindle may be
lifted out together, exposing the whole
interior of the flush-pot, which may thus
be given a daily cleansing and kept in as
good order as any other iron vessel in the
kitchen.

The theory of the success of this apparatus is very simple. There is absolutely nothing running through the waste-
pipe except during the moment when the
flush-pot is being discharged, and then
the whole mass flows with such force as
to carry every thing with it.

At my own house, having occasion to
inspect the main drain (diameter three
inches), I found that neither a copiously
supplied water-closet nor a bath-tub had
such flushing effect as had the discharge

of the flush-pot in the kitchen. Its flow filled the drain more than half full with a stream of good velocity.

In the first application of the flush-pot to pantry sinks. it was given about the same capacity with that of the kitchen sink. As, however, it is desirable to fill the pantry sink for washing dishes, it became necessary to waste the large volume needed to fill the flush-pot. To avoid this its capacity has now been reduced to about one gallon, which is enough to insure a good flow from the ordinary accumulated drippings of the sink. When the sink itself is filled, its contents as well as those of the flush-pot constitute an abundant flushing volume, the strainer being sufficiently open to allow a rapid flow.

NOTE.—This device has not come into general use. It requires some personal attention, and this cannot be depended upon with changing cooks. I find it most satisfactory in my own house for both kitchen and pantry.

Experience has shown, however, that the frequent stoppage of sink-traps and waste-pipes of large size is chiefly due to the fact that the small stream flowing through them has not sufficient force to keep them clean.

After use for a certain time, the gradual accumulation of congealed grease on the walls of the pipe closes the passage almost entirely. We have, then, not a direct channel with smooth metallic walls, but a tortuous and irregular one, winding through a mass of deposit. The detachment of parts of the decomposing mass may at any time cause complete obstruction. Then follows the use of potash. to " cut " a passage through it, or of hot water to melt the grease. This being troublesome, a costly and elaborate " grease-trap " is sometimes used to hold back the offending material. All goes well for a time, until the trap itself becomes filled, save for the tortuous waterway above described. In time the grease-trap, which servants are sure to neglect, as well as the pipe it was intended to protect, is cleaned out by the plumber, and a fresh start is taken.

The real remedy is to make the trap and pipe so small that the natural flow from the sink will keep it always clean. I have not dared, in the face of the universal belief that a pipe that is liable to become obstructed should therefore be made large, to advise the use of kitchen wastes less than one and a quarter inches in diameter. Two inches is generally thought to be as small as is safe. It is my own conviction, however, that it would be safer to make the diameter *less than one inch*.

This idea has had curious confirmation in a case that has just come to my notice. In 1885 I put in a kitchen waste-pipe, of lead, at Princeton, N. J., which was one and a half inches in diameter. It ran directly down through the floor and then continued horizontally (approximately) across the kitchen about twenty feet. In

FIG. 9.

time the floor timbers rotted and the springing of the floor flattened the pipe. After eight or nine years the house was converted into " Evelyn College," with some thirty students, the family, and servants—all living in the building, and all served from this single kitchen. In 1894 an inspection of tbe plumbing was called for.

It was found that this waste-pipe had—surely some years before—been jammed almost flat. A section of it was cut out, and its end was smoothed to make a type from which Figure 9 was printed. It seems incredible, but through this small slit the kitchen-sink wastes of a household of about forty persons flowed for some years without anyone knowing that its original water-way had been contracted. The area of the cross-section of this flattened pipe was not more than *three-tenths* of an inch. The chances for good service would have been much better with a circular section half an inch in diameter, and I verily believe that a waste-pipe of this size, protected by a strainer with quarter-inch holes, would be safer than a larger one. We often find the winding channel through the greasy deposits of a 2-inch pipe to be less than this.

The world seems to have a curious infatuation in favor of some sort of "grease-trap." In Chicago its use is compulsory. As generally constructed, it is a sanitary nuisance, and I know of only one device of the sort that I would use. This is the "Eduction" grease-trap, made by Flushtank Company, of Richmond, Ind. Its deposits are torn out and washed away by a frequent strong flush. For my own use, I should prefer a very small waste-pipe, protected by a good strainer and a Puro trap.

CHAPTER XXII.

OVERFLOWS, intended for the safe removal of surplus water from bath-tubs, wash-bowls, etc., are necessarily on the house side of the trap. They are practically never reached by a strong flushing stream, and their walls accumulate filth and slime to a degree that would hardly be believed. They constitute the nastiest element of modern house-drain-age of the better order. Perhaps they are not a serious source of *danger*, but they are, more often than any other part of the plumbing work, except the urinal, the source of the offensive drain-smell so often observed on first coming into a house from the fresh air.

In the stationary wash-basin as at present arranged, there seems to be no easy way to get over the difficulty, a difficulty which of itself should be a sufficient reason for excluding these fixtures from sleeping-rooms. The basin overflow is objectionable for substantially the same reason that the bath-tub overflow is objectionable, though perhaps to a slighter degree owing to the smaller surface exposed to the accumulation of deposits.*

The concealed overflow of the bath-tub may, fortunately, be dispensed with, and in this case the difficulty inseparable from the arrangement may be obviated. It will, perhaps, be instructive to illustrate by a diagram the reason why the usual hidden overflow is so objectionable.

In this cut, *A* is the waste-pipe at the bottom of the tub, by which its contents

---

* See the " Note " at the end of Chapter XIX.

are discharged on the withdrawal of the plug. *B* is the overflow pipe, its connection with the tub being through a perforated screen. *C* is the trap by which

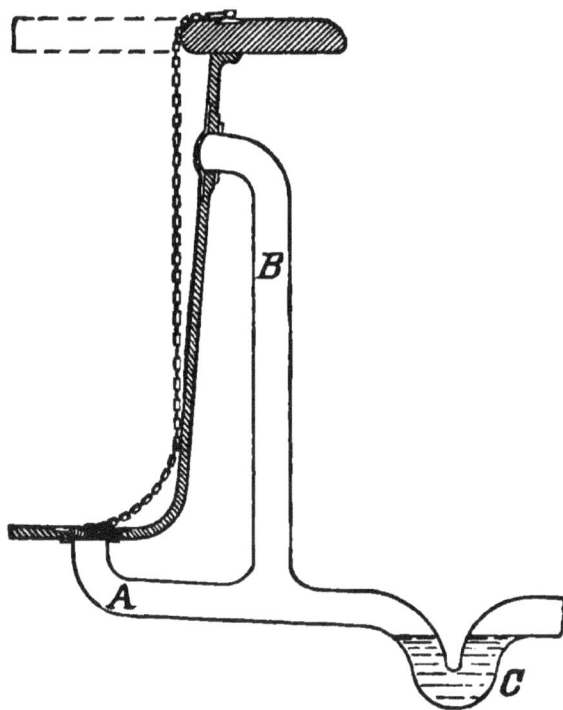

FIG. 10.—HIDDEN OVERFLOW OF BATH.

the waste-pipe is shut off from the drainage system, and which has incidentally the effect of retarding the flow of water through the waste-pipe. If we suppose

the tub to be filled to the level of the
overflow and its waste-plug to be removed,
the water will immediately rise in the
overflow pipe to very nearly its height in
the tub. It is of course impregnated with
the impurities of the water in the bath.
Furthermore, the lighter particles of
organic matter flowing through the waste
will, some of them, rise by their levity
into the overflow pipe. The water rushes
up into this pipe with much force, but it
descends only very slowly as the level in
the bath descends, so that at each operation there is a tendency to deposit adhe-
sive matters to its walls. What is so deposited decomposes and escapes little by
little in a gaseous form through the perforated screen into the air of the room.
There is a free circulation of air through
the pipe when the plug is out. The
amount of these decomposing matters is somewhat increased, though prob-

ably not very much, by floating particles
passing through the screen when the
overflow is performing its legitimate func-
tion.

This is the simplest statement of the
proposition, and this is perhaps the least
objectionable form of hidden overflow.
Where the waste-pipe is closed at the
bottom of the overflow by a plug or valve
attached to a spindle rising through the
overflow-pipe—a very favorite device
with some plumbers, and already described
in connection with wash-bowls—the difficulty is in every way aggravated and the
amount of fouled surface is much increased. The inherent defect here illustrated attaches to every overflow of this
general character connected with any part
of the plumbing work.

In the case of a bath-tub it may very
easily be avoided, as shown in the next
diagram, by doing away entirely with the

overflow-pipe *B* and its perforated screen,
and using for the closure of the waste-
outlet *A*, as a substitute for the ordinary
plug, a pipe fitting into the outlet and

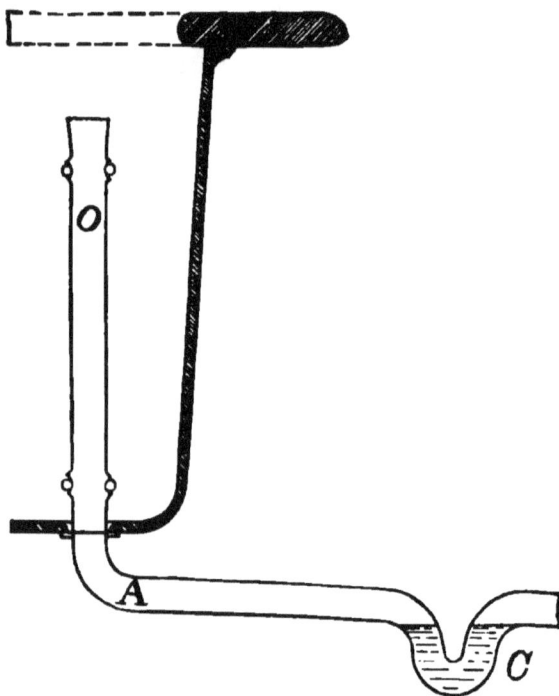

FIG. 11.—STANDING OVERFLOW AND PLUG FOR BATH.

rising to the height desired for the water
in the bath.   If the upper end of this
pipe be given a trumpet-shaped opening,
its capacity will be increased.

Stop cocks need no special notice in this book, except in connection with bath-tubs. Most, if not all, of the English earthenware bath-tubs imported into this country, and many even of the planished copper, and enameled iron tubs made here, are furnished with an ingenious device for delivering the supply near the bottom of the tub in such a manner as to mix the hot and cold water at the delivery and to admit the supply with little noise. The last may be an advantage. The first may be perfectly accomplished by delivering the hot and cold water through a single nozzle at the top of the tub in a convenient position for drawing water for other uses.

There are doubtless many cases where the bottom delivery of the supply may be free from sanitary objections, but they are fewer than would be supposed, and it seems strange that the frequent serious

objection to the arrangement should have
been so generally overlooked. This bottom delivery is substantially a cock for
drawing water, and all who use such
cocks for filling wash-bowls must have
noticed a frequent indraft of air when the
cock is open. Sometimes when water is
being drawn from the lower part of the
supply-pipe, the head in the upper part
is annihilated, and if a cock is opened the
water falls in the supply-pipe, air rushing
in to take its place.

The indraft of *air* is not of much consequence, but the indraft of a pipeful of
dirty water from a bath-tub does not suggest a pleasant modification of the quality
of the water supply of the house. In
this case, as in many others, an apparent
mechanical improvement, securing only
incidental benefits, should be discouraged.
In my judgment the only perfectly safe
and satisfactory arrangement for baths

thus far devised is one by which the water is drawn through a faucet above the water-line, and by which the outlet is closed by a stand-pipe serving as the *only* overflow of the tub.

Laundry trays, as they are now almost universally arranged, are hardly to be regarded as a conspicuous element of the sanitary works of a house. There are few cases in which we find anything about them that is seriously objectionable. With them, as with sinks, water-closets, and wash-basins, it is best to avoid all unnecessary carpentry. It is, of course, best that they should be made of some other material than wood—either slate, soapstone, cement, or earthenware. Earthenware tubs, supported on galvanized iron legs and surrounded by a simple border of hard wood, seem to ask for no improvement.

CHAPTER XXIII.

SIMPLICITY in house-drainage, and a marked contrast to the multiplication and complication so often found in the better class of buildings, are illustrated in the case of a very fine and costly house, the plumbing of which I am now superintending.

It has in the basement one kitchen-sink with the flush-pot, one servants' water-closet, and four laundry-tubs. The main soil-pipe runs under the basement floor near both of these; it is of extra-heavy iron, with the joints leaded, and tested, under water pressure, to absolute tightness. It is then, so far as it lies below the floor, completely

encased in Portland cement mortar, and
this, again, in well made concrete; it
turns up near the laundry-tubs, and near
the ceiling it receives a branch pipe coming from a lavatory on the floor, twenty-
five feet away; it then passes through
the floor and receives the waste of the
flush-pot of the pantry-sink; rising to the
ceiling, it receives the waste of a bath-
tub and wash-stand on one side, and on
the other the waste of a Dececo water-
closet and wash-stand; passing through
the next floor, it receives the wastes of
the fixtures in the servants' bath-room—
a straight hopper closet, a bath-tub, and a
wash-stand; above the ceiling of that
room, its four inches size is increased to
six inches, and it passes with this larger
diameter a short distance through the
roof, its top being closed by a large wire
basket inserted in the hub of the six-inch
pipe; the branch pipe under the ceiling

of the cellar is connected with a Dececo closet and a wash-stand in the lavatory, and is continued up, without other con- nections, to its increaser and a six-inch top joint through the roof.

This is the full complement of the drainage appliances which, in accordance with modern ideas, it was thought neces- sary and wise to introduce into a house which, even five years ago * would have had twice as many closets and baths, and at least four times as many wash-basins, to say nothing of two or three urinals and one or two house-maid's sinks. The whole cost of the work to be done, including all hot and cold water-supply, and the outside connection with the sewer of six roof-water conductors, is just about one thousand dollars. Under the old method, supposing the same material and workmanship to be used,

* 1880.

and considering the long lateral waste-
pipes and hot and cold water and circulation pipes of the different baths and
basins, the cost would hardly have been
less than twenty-five hundred dollars.
The saving of cost effected is, in my
judgment, of much less consequence than
the simplicity secured.

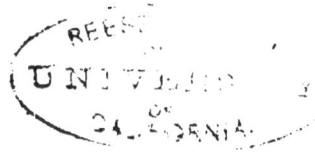

CHAPTER XXIV.

OWNERS, ARCHITECTS AND PLUMBERS.

IT has already been said that the character and arrangement of the plumbing work of modern houses are much more controlled by plumbers than by any body else. This is generally quite as true with reference to houses built under the advice and direction of architects as in other cases.

There are a few architects who take a lively interest in sanitary works, but they are very few. The fact is, that an architect is, generally, either disqualified or disinclined, by nature or by training, for the minute attention to hidden details which the proper control of house drainage would require of him. It is not to be expected that a man who has made himself a safe guide and leader of public taste, who has acquired a mastery of the engineering problems involved in safe and economical construction, and who has learned how to contract a client's desires to the requirements of space and of price, should have been able to keep himself *au courant* with the rapidly growing improvements of an art, which, twenty years ago, was only a trade.

As a matter of fact, the architect rarely knows or cares any thing about the plumbing of a house beyond the selection of the spaces to be given to its fixtures. He gives the plumber a rough idea as to where certain fixtures are to be placed, and as to their general style. Beyond that—I am not speaking of all architects, but of architects as a class—he is generally indifferent to the whole business.

The result is—as it may be. Some plumbers are capable of writing good specifications and some are not. It by no means always happens that the best plumbers and the best architects work in conjunction.

I write more feelingly on this subject, because I have recently, in more than one instance, on being called to take charge of the drainage of very fine houses already in course of construction, had the plumbing specifications submitted to me. They happen in all cases to have been written for places where the Board of Health has established no formula for the work and has furnished no printed blanks. Not one of them would have done fair credit to an intelligent plumber of the year of our Lord, 1870. In every case an unnecessary amount of work would have been done and much of it would have been unsafe when it was done. The

question often comes up, in the practice
of an engineer of sanitary drainage, as to
whether or not to change such work as
these specifications provided for, it being
already constructed and in operation.
No one who knew the first rudiments of
the business, would think of introducing
it into a new building.

In one case, I was called to a house so
near completion that all of its main pipes
had been put in place, and a wilderness
of pipes they were. I recommended that
they be removed, all but one of them,
and that a fresh start be taken. The
owner referred the question, as owners so
often will, to the arbitration of dollars
and cents. An estimate, made by the
plumber who had done the work, showed
that it would be somewhat cheaper to
wipe it all out, begin again, and follow
my plans, than to finish what he had
begun, according to his own.

The criticism here is in no sense against the plumber nor against the architect, but against the *system*. If an architect tells his client that he prefers to have an expert to take charge of the plumbing, the client may object to paying an extra fee or, as it seems to him, two fees for the same work. The architect, of course, can not be expected to pay for the extra services, so he follows the old course of turning the matter over to his plumber, and, very properly, justifies himself with the sufficient plea of Usage. As an incidental result, the owner generally pays considerably more for the tolerable work that he gets than he would pay, fees included, for the more perfect work that he might get.

Another point in this connection is of great interest. It is being actively discussed by the better plumbers and by the sanitary journals. It is that plumbing work should, at least, never be done at second hand. That is, that the whole job should not be given to a contractor, allowing him, in his turn, to contract out the plumbing work. Indeed, this results so generally in positive injury to the health of the people that it would seem a proper subject for legislative prohibition.

Naturally, all that the building contractor is responsible for is, that water shall flow and waste shall run, without bursting the pipes or destroying the ceilings, until after the house has been accepted and taken off his hands. His only connection with the enterprise is a business connection and his only motive is his business interest. He builds the house to make money on it, and the more cheaply he can get the architect's requirements for water-closets, baths, tubs, sinks, etc., complied with, the more money he will make.

While this course of sub-letting is bad enough in the case of houses built for immediate sale, or in the case of tenement houses, it is in a certain way explicable ; but that any man who is building a house for his own occupation, should permit this vital element of its safety as a habitation, to become the object of dicker and flint-skinning between two men who care only to make what money they can out of it, is most incomprehensible.

If it is suspected that what is here written is written with a view to the pecuniary interest of my own profession, with its rapidly growing membership, I confess at once and most frankly that the suspicion is well founded. I should not have taken the trouble to write this book for the questionable compensation it offers in the way of copyright. I have written it because I think it may advance the interests of engineers. If that should be

thought an improper or a selfish motive,
let me say that I trust it will also, and in
much greater degree, advance the interests of all who live in houses which have
or which need drainage works of any
kind. That it will also benefit the better
class of plumbers, ought not to be
doubtful.

CHAPTER XXV.

THOSE who live in isolated houses, in the country and in the smaller towns, generally solve the whole problem of sewage disposal by the simple formula, "Out of sight, out of mind." The universal remedy is a hole in the ground into which every thing in the shape of liquid wastes is delivered. The only criticism that most people apply to the apparatus relates to the frequency with which it requires to be emptied and to the cost of the operation.

What goes on in the cesspool, and in the ground about it, is entirely unheeded. So long as the overflow is disposed of,

the devil may care what the hidden processes of disposal are. Unfortunately,
the devil does care, and much of the
worst work that he effects, in his assault
on the health of his unwitting subjects,
has its starting point in the common
cesspool. The worst sewer in the world
is rarely so bad as the usual cesspool.
In comparison with it, a sewer that would
be regarded as very foul is purity itself.

The cesspool works its injury chiefly
in three directions:

A. It holds an accumulation of filth
of the worst character, in a state of
active putrefaction, giving off gases produced by a decomposition that takes
place under conditions of the worst sort.
These gases can not be entirely sup-
pressed while a channel remains open
for the admission of liquid to the cavern.
Unless delivered into the atmosphere,
which they inevitably taint, they find

their way back more or less directly into the drainage system of the house.

B. In so far as the walls of the cesspools and the surrounding earth are permeable to their passage these gases pollute and poison the ground air, and if near the house, this may find its way from the earth into the cellar.

C. Its liquid leachings and ooze, enriched with the soluble productions of a fatal putrefaction, travel through porous soils, through gravel or sand streaks in heavy ground, and through fissures in rocks to pollute the water of wells and springs even at a considerable distance.

If the fraternity of sanitarians throughout the world are in accord on any one point, it is that the common cesspool, even under the best conditions, is absolutely inadmissible. Surely no householder having the least regard for the health of his own family, or for that of his neighbors, once realizing the inevitable condition, will tolerate its perpetuation. Fortunately it is no longer necessary that, even in connection with the simplest and poorest house, this dangerous nuisance should exist. The sanitary fraternity are also in full accord as to the general principles of an improved mode of disposal.

What concerns us, in this connection, is a proper means for getting rid of the organic refuse and filth contained in the liquid wastes—including those of the water-closet and kitchen sink—of isolated houses which have at least a small area of land connected with them.

The principle on which the best disposal thus far devised is based is this :

The aerated upper layers of the soil are a universal destroyer of whatever organic impurities may be deposited on the surface or intermixed with the earth. Dead organic matter is completely destroyed by natural processes of oxidation and nitrification, wherever, under natural conditions, it is exposed directly or indirectly to the action of the atmosphere in the soil. The agents of destruction in this case are the bacteria of decomposition, those all-pervading micro-organisms, whose increase, growth, and life involve the combination of atmospheric oxygen with the elements of organized matter, a combination always set up under conditions favorable to the growth of these bacteria. Aeration is a necessary condition of the process.

When urine or other foul liquid is thrown on to the surface of the ground, it sinks into it and deposits on the surfaces of its particles the organic impurities which it contains. Its water, descending still further, has become essentially purified. As the water sinks away, air enters

to take its place, and every particle of
soil that is coated with the waste matters
is surrounded with air rich in oxygen.
The bacteria, present in all fertile soils,
combine the two—much as the respira-
tory process of higher organisms com-
bines the oxygen inhaled by the lungs
with effete matters of venous blood.

This is a slight and popular statement
of the results of recent scientific investigation.   The process is a continuous one.
The destruction of one dose of wastes
being accomplished the destroyers are
ever eager for more.   For their successful and continued operation it is only
necessary that the waste matters should
be supplied and that air should be furnished alternately.   This is the principle
on which is based the success of the system of sewage purification by broad
irrigation and by intermittent downward
filtration.   It is as effective in taking care

of the waste of a city as of that of a single house, the difference being only a difference of degree.

In the case under consideration, the disposal of the wastes of a single dwelling, generally on land very near the house, may be accomplished in a hidden and inodorous manner. The sanitary result would be complete if they were delivered intermittently over the surface of a lawn. Other considerations sometimes make it important that sewage be kept out of sight and that the distribution of the wastes be beneath the surface.

The Rev. Henry Moule, the inventor of the Earth-Closet, seeking means for disposing of liquid wastes of the household, for which his closet is not suited, laid agricultural drain tiles along the foot of a grape-vine trellis, leaving their joints open, so that the slops reaching the drain could escape freely into the ground. The influence of this subsoil irrigation on the growth of the vines and of grass was very marked and the disposal was complete and continuous. Later (1868) Mr. Rogers Field adopted a modification of the same device in disposing of the liquid wastes of cottages at Shenfield, Essex, England. He added the important improvement of a flush-tank in which the liquids were retained until it became full, the whole volume being then rapidly discharged into the tiles.

This was a great advance. It secured the uniform distribution of the liquid throughout the whole length of the drains and gave ample time, while the tank was refilling, for the water to settle completely away, allowing air to enter and complete the destruction of the impurities before the next discharge.

The writer, in connection with his residence at Newport, adopted Mr. Moule's device in 1869, as soon as it was made known, and added Mr. Field's improvement soon after the publication of his process. At first, a little difficulty resulted from the delivery of grease and flocculent matter into the tiles, leading to the obstruction of their joints. Early in the experiment, it usually became necessary as often as once a year to lift the tiles and wash out their accumulations. This led to the interposition of a settling basin to retain all of these coarser matters. Later the settling basin was placed on the house side of the flush-tank to prevent the disturbance of the deposits in the former by the vigorous discharge of the latter.

The house was provided with earth closets, but everything not deposited in these was delivered to the drains, which were laid under a lawn adjoining the best rooms of the house and a piazza that was much used. The net work of drains began not more than fifteen feet from the edge of the piazza and the furthest corner of the section was not more than fifty feet away. The system worked most satisfactorily for eleven years, when the construction of a sewer in the adjoining street made it no longer necessary to use it.

During the whole period of the experiment, the condition of the tiles and of the soil immediately adjoining them was frequently examined. If the opening were made immediately after the discharge of the tank, the earth near the drains had a decided odor, but all other times, even close to the open joints from which the discharge had been copious, there was absolutely nothing to suggest impurity. A handful of earth which a few hours before had been saturated with the foul liquid had no other appearance or odor than that taken at the same depth from a part of the lawn entirely beyond the influence of the system.

This method of disposing of waste liquids has been used by many engineers and on hundreds of places in different parts of the country. So far as known, wherever the details of the system have been properly regulated, the result has been entirely satisfactory.

In 1876, in connection with the sewerage of Lenox, Mass., there being no other available means of outlet, a large flushtank was built at the upper corner of a field near the village, and 10,000 feet of absorption drains were laid. This system was used until another outlet was provided in 1892. Whenever it was properly cared for it was completely successful.

At the hotel at Bryn Mawr, Pa., a system even larger than that at Lenox has been in successful use since 1881. Here, the sewage had formerly been delivered into a brook which ran through private property. The change was made to avoid legal proceedings threatened by the adjoining owner. A little difficulty occurred at first from an unequal distribution of the flow and because of the steep grade necessarily given to some of the drains. These defects were remedied and the working of the whole system has since been satisfactory.

At the Woman's Prison at Sherburn, Mass., where about 30,000 gallons of sewage were delivered per day, the conditions were extremely difficult. Only a small area of land at considerable elevation could be reached, and this was of a very unfavorable character, The system here was on the whole successful, though owing, as is believed, to sluggish drainage through the heavy subsoil, the amount of drain tile used, 20,000 feet, at times proved inadequate to the work. Some years ago a public sewer was constructed into which the prison drain now delivers.

There are few individual houses so far separated from each other that their best relief may not come by public sewerage, which have not sufficient ground about them for the successful application of this system.

Although it was not the intention to enter in this volume so far into the practical technicalities of work as to use any considerable number of diagrams in illustration, it is thought that, as the system under discussion, though largely used in a few Eastern neighborhoods, is still little known over the country generally, it will be proper to make an exception here, and to reproduce from a descriptive article, published in the

American Architect for March, 1892, the diagrams accompanying it.

Figure 12 shows the construction of the double-chamber tank. The settling-chamber *A* is a small, round cistern with a wide throat—not less than

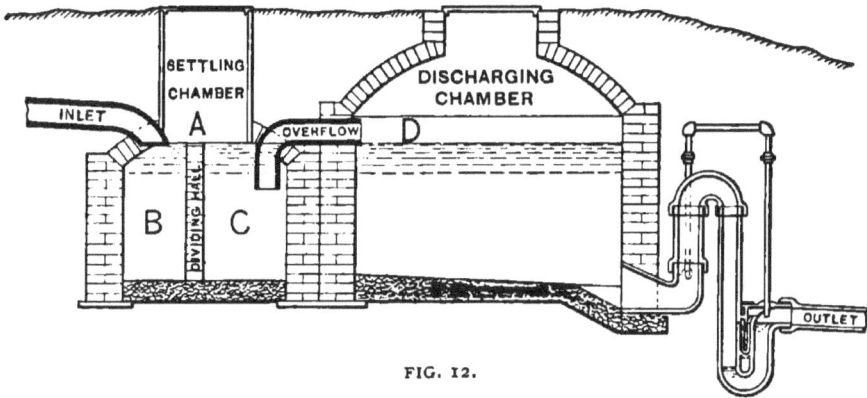

FIG. 12.

eighteen inches diameter—to facilitate the removal of its scum and deposit. It receives sewage from a pipe turned down through the dome and barely trapped against the return of air—if deeply trapped, grease accumulates and obstructs the drain ; with this slight

trap, the flow from an ordinary house
suffices to keep it free. It overflows
through a deeply-trapped pipe into the
discharging-chamber *D.* It is divided by
a wall into two chambers, the top of the
wall being just at the overflow line.
The compartment *B,* on the inlet side,
has its water considerably agitated by
the inflow. Before the dividing-wall was
adopted, this agitation was communi-
cated to the contents of the whole chamber, and flocculent matters, which would
settle to the bottom or float to the top in
still water, were carried over by the current into the discharging-chamber. This
agitation is now confined to the compartment *B,* from which the liquid portion
flows to the compartment *C* in a thin
sheet over the top of the wall, in such a
manner as not to disturb the contents of
this second compartment, allowing flocculent solids to settle quietly to its bot-

tom. Under some circumstances, perhaps due to a higher temperature in the
sewage, and this to its larger amount, the
decomposition of the sediment and of
the scum is sufficiently active to prevent
accumulation to an injurious amount.
In such cases, the settling-chamber need
never be cleaned. This is not to be
depended upon without occasional
inspection. In the majority of cases it is
necessary every few months to bail out
the chamber and get rid of its accumulations, which should be buried or dug into
the ground at once.

The liquid overflow from the settling-
chamber *A* to the discharging-chamber
D represents practically the full amount
of sewage brought down by the drain.
The discharging-chamber should be
made large enough to hold the product
of at least twelve hours; there is no
objection to its retaining twenty-four

hours' supply,—a longer retention would
lead to too much putrefaction.   This
chamber is furnished with an automatic
siphon ; the one shown is what is known

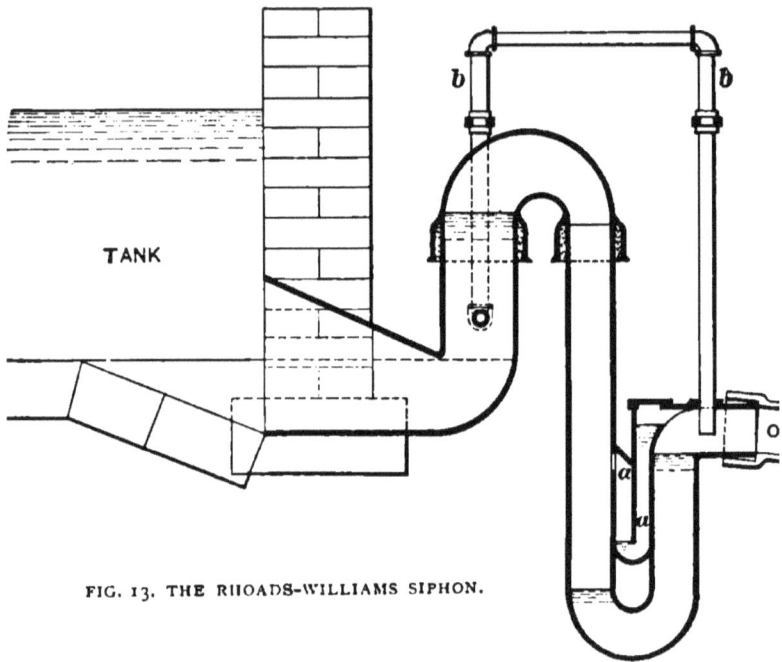

FIG. 13. THE RHOADS-WILLIAMS SIPHON.

as the Rhoads-Williams siphon.   Its details are shown in Figure 13.

It depends for its action on the sudden
releasing of compressed air contained
between the inflow from the tank and

the deep trap near the outlet. When the pressure is sufficient to force the water in the blow-off trap *a, a,* to the bottom of this trap, the air pressure is released, and the head of water, which it had held in the tank, forces a full flow into the siphon and brings it rapidly into action. Air is introduced for the breaking of the siphon after the main flow has ceased by the admission of air from the drain through the pipe *b, b.* These siphons are sold by flush-tank dealers.

The siphon is located entirely outside of the tank. This obviates the serious fouling of the siphon itself, which has always been a source of difficulty when it was placed within the tank. Its opening into the tank is funnel-shaped so as to take the flow rapidly, and to make sure that there will be no obstruction from such minor solid matters as the sewage may contain. It is not ordinarily found

necessary to clean out the discharging-
chamber, matters which would otherwise
accumulate within it being held back in
the settling-chamber.

The outflow is a slightly putrid sewage
containing more or less fine flocculent
matter, not enough to interfere with the
proper action of 2-inch absorption-drains.
These absorption-drains may be placed
at a greater or less distance from the
tank as the tank may be at a greater or
less distance from the house. They are
made of ordinary round 2-inch tile in
one-foot lengths, laid in earthenware
gutters, their joints being open about
one-fourth inch, and being protected
against the entrance of earth by the
loose-fitting cap laid on the top. The
gutters and caps are of larger radius
than the outside of the tile, so that practically the whole joint is available for the
escape of sewage into the ground. The

surface of the gutter on which the tile is laid should be ten inches below the finished surface of the ground. In a reasonably porous surface-loam, it will suffice to have one foot of tile for each gallon of the contents of the discharging-chamber. The tile, caps, and gutters are shown in Figure 28.

If the soil is heavier, the length must be increased. An impervious clay is not well suited, under any circumstances, for this use, but where nothing else is available there should be at least three feet of tile per gallon. The tile may be one continuous line, or a number of shorter lines, connected with the 4-inch main leading from the flush-tank. The tiles need not be more than three feet apart, though twice this distance is not unusual. In fact, the system is in this respect a very flexible one, and can be adapted to land of any shape or inclination. The fall of the main line from the tank to the absorption-drains should not, especially after coming within 20 feet of the first line of tiles, have a fall of more than 4 inches to 100 feet. Its joints should be cemented, and the branches for connection with the tile lines should come out from the bottom of the tile, not from the middle, as is usual with branches of vitrified pipe. Special pieces for this purpose are to be obtained from the dealers. The absorption-lines themselves should have a fall of not more than 2 inches per 100 feet : more than this gives a tendency to an accumulation of sewage at the far end of the line, and if the line is long, to a breaking out of the sewage at the surface.

Figures 14, 15, and 16 show three different methods of applying this system, according to variations of the ground.

In each case the dotted lines are con-
tour-lines showing differences of elevation of one foot.

In Figure 14, the flush-tank *A* receives
its sewage from the sink-drain leading
from a corner of the house. Its dis-

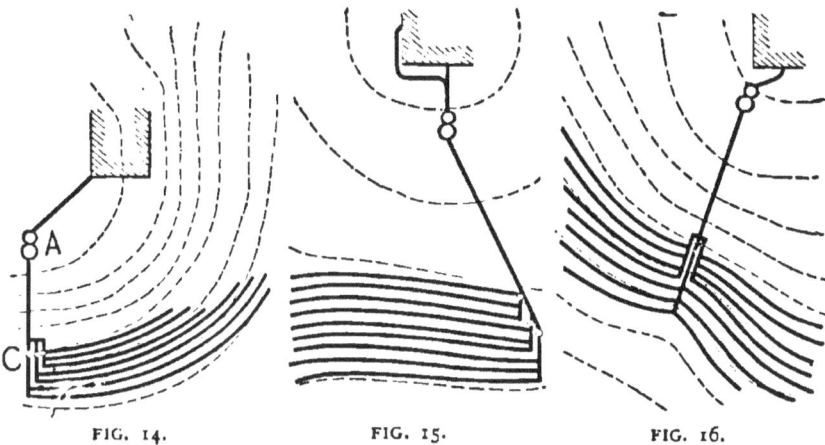

FIG. 14.   FIG. 15.   FIG. 16.

charge is through a direct line to the
point *c*, starting at the tank at a depth of
three or four feet below the surface and
coming to within a foot of the surface at
the point *c*. At the point *c* the main
drain is turned at an angle and has three
different outlets, to be used, one at a

time, in alternation. The first one communicates with two parallel drains at the
bottom of the field, these having a sufficient combined length to receive the
whole discharge of the tank. The second one runs parallel to the first, to
two absorption drains corresponding with
the first two. The third communicates
with the other system of three parallel
lines, which are shorter, having about the
same aggregate length as the two of the
other systems. They are carried around
nearly parallel to the contours to secure
the requisite slight fall.

In Figure 15 the flush-tank is fed by
two drains from the house and connects,
as shown, with three series of three
drains each. The land is much more
nearly level, and the nine absorption-
drains are in ground having a total fall
of only one foot.

In Figure 16 the flush-tank is fed by a

single drain, not straight, and connects with its alternating gate. There are two series of drains, four on one side of the field and four on another, while a third line connects with the series of three shorter drains on each side of the medial line.

The flush-tank may be placed in any position and at any distance from the house, and the field may be at any distance from the flush-tank to which a proper fall can be obtained.

One modification of this system, which will obviate much of the difficulty by requiring a greatly reduced retention of solid matters, where indeed a coarse screen may, in some cases, be made to hold back all that is necessary to retain, consists in the use of larger absorption-tiles, say 4-inch or 6-inch, with open joints, fully one-half inch wide and laid in coarse gravel or other very open

material just under the surface of the
ground in two, or better, three series,
each of which has sufficient capacity in
its pipes to receive the entire contents of
the tank at each discharge. The discharge being at intervals of from twelve
to twenty-four hours, the liquid sewage
with its soluble and its finer suspended
impurities will have ample time to leak
out into the soil; and during the period
of intermission, while the other two
series of drains are being used, worms,
beetles, and other insects will consume,
or decomposition will destroy, the relatively small amount of deposited matter,
which, however, might be sufficient to
obstruct 2-inch tiles.

A still further modification consists in
making the drains of similar large tiles
of "horseshoe" shape, laid in a trench
filled with coarse gravel or broken
stone. The capacity of the tiles and of

the voids among the stones in each series should be sufficient to receive more than the full contents of the tank.

Cross-sections of 4-inch horseshoe tiles, laid in trenches filled with stone or gravel, are shown in Figures 17, 18, and 19.

In Figure 17, the ground is supposed to be reasonably absorptive, like garden mold. In Figure 18, the natural soil is very heavy and non-absorptive, and is level or nearly so. It is thoroughly underdrained and is covered with sand or gravel in low ridges, being deeper at the absorption-drains than midway between them. The purification takes place entirely in this porous and well-aerated surface, the clarified water sinking into the drained ground below. In Figure 19, the land is non-absorptive and has a decided slope. It should be well underdrained with tiles running up and down the slope. The surface is covered with sand or gravel, and is divided into sections by banks of clay (under the sand) at the foot of each section. The sewage is delivered into a horseshoe tile, with

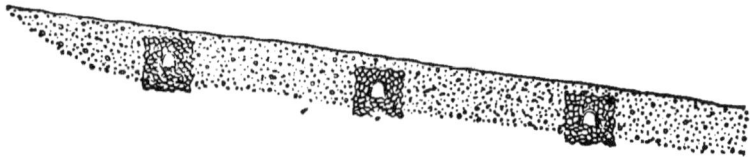

FIG. 17.

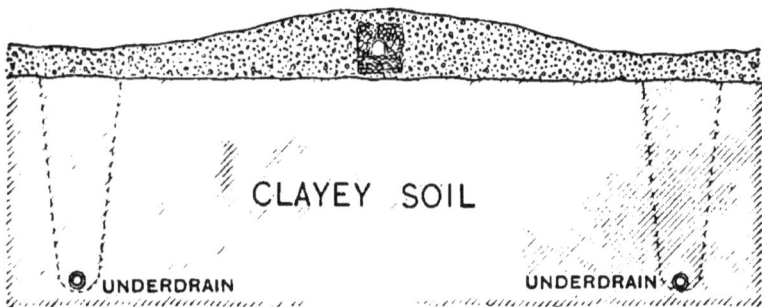

CLAYEY SOIL

UNDERDRAIN   UNDERDRAIN

FIG. 18.

broken stone, at the upper side of each section, and is purified (and its water is absorbed) before it reaches the clay bank below, or is held by this bank until it is disposed of.

This question is often raised with

reference to the sub-surface system : " If the dire results of the use of the ordinary leaching cesspool are so serious, why is it not just as bad to discharge the same materials into the ground by leaching

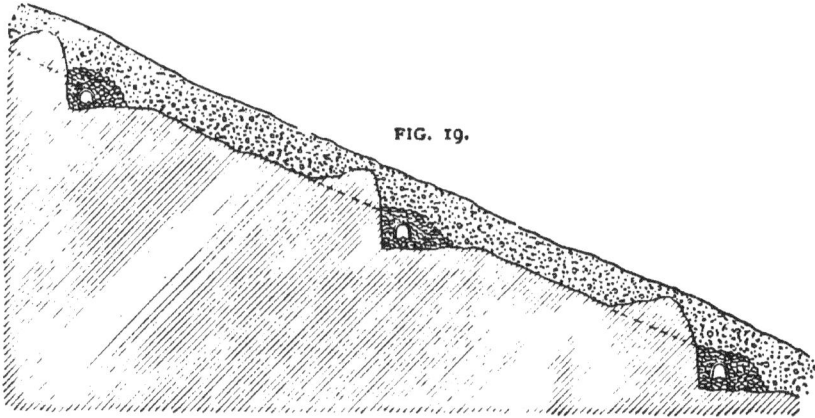

FIG. 19.

drains ? " The difference is radical. In the case of the cesspool, the leaching takes place almost entirely at such a considerable distance from the surface that the exclusion of air makes suitable bacterial action impossible, while the delivery through drains laid immediately under the sod is into aerated ground which is teeming with bacteria.

In the system described above, a certain amount of putrefaction is inevitable, and putrefaction is always objectionable. When once its fæcal matter is submerged, so as to prevent the exhalation of its odors into the atmosphere, *fresh* sewage is entirely inoffensive to the smell. It is simply so much dirty water. It might be thrown in considerable quantities on to a grass plat without other objection than that which would attach to the sight of its solid portions. Unless thrown so frequently and in such quantities on the same spot as to saturate the ground, exclude the air, and so prevent proper bacterial action, it would produce no odor whatever, barring the trifling odor of exposed fæces. The same sewage retained for two or three days in a barrel, vault, or cesspool would enter into a state of offensive decomposition, and would constitute a nuisance of a serious character.

The real problem in the disposal of household sewage, and indeed of the sewage of large institutions, and even of towns, is to bring it into contact with a suitable oxidizing medium while in so fresh a condition as not to be offensive, and to make sure that the drains and other appliances through which it passes shall at no time become offensive. Offensive odor is, of course, chiefly to be guarded against; but offense to the sight would also constitute an important objection. No method has as yet been devised by which the whole process could be appropriately carried on on the front lawn of a dwelling-house. We have, however, arrived at a point where, even in a reasonably secluded back-yard, all of the conditions may be satisfied. Setting aside exceptional cases, where only the sub-surface system with a double-chambered flush-tank would be acceptable, the work may be done in a great majority of instances by a system of surface disposal or of sub-surface disposal with large pipes, with an automatic flush-tank which may at all times be thoroughly aerated and which may be cleansed with sufficient frequency to prevent odor from its sliming. Especially in connection with such a flush-tank can the ventilation of the main drain and soil-pipes be maintained with such completeness as to prevent the accumulation of odors in them. The objection to placing a flush-tank of this character quite near to a house is rather fancied than real.

The nearer it is and the more constantly subjected to inspection, the greater will be the certainty of keeping it always in good condition. On the other hand, the farther from the house, the more will the solid parts of the sewage be broken up in transit, the less impera-

tive will be the need for scrupulous cleanliness, and the less the care required ;
civilized living, however, requires great
care in all such matters. Civilization is
only lately beginning to take cognizance
of this requirement, and the public at
large is still disposed to like best a system which calls no attention to itself,
and of which the objectionable features
are so hidden as to be easily forgotten.
Even under their best development, such
systems belong to the "out-of-sight-out-
of-mind" class.

Accessibility and exposure to constant
inspection, now universal in the best
plumbing practice of the day, are equally
desirable in the case of apparatus for the
disposal of sewage outside of the house.
The cesspool, the vault, and the double-
chambered flush-tank will soon have become things of the past, among those
who care for good sanitary conditions.

Their places will be taken by some device having the essential features of the
system described below, combining efficiency with the possibility of, and the demand for, perfect cleanliness. This being accomplished, the "horrid drains"
will soon cease to exist, or to be thought
about.

It is not to be supposed that this pre-
cise manner of applying the system will
long remain the accepted one. But the
principle of the arrangement seems to
embody all the elements of permanence.
This principle may be thus stated :

Deliver the sewage as soon as produced, through thoroughly flushed and
ventilated pipes and drains, to a point
outside of the house ; hold back its
coarser substances by some form of
screen which will allow everything to
pass that can, in an unobjectionable way,
be disposed of with the sewage ; accumu-

late the regular flow through the drain
in a receptacle large enough to hold the
supply of a few hours, or of a day, as the
case may be ; the receptacle becoming
full, discharge its contents automatically,
rapidly, and completely, on to the surface
of the ground, or into drains immediately below the surface, for its final,
complete, and inoffensive disposal ;
arrange the screen and tank in such a
way that they may be kept in a cleanly
condition with little labor and without
requiring the constant supervision and
nagging of the master of the house.
The appliances shown in the Figures 20
to 27, and the method of working
described in connection therewith, seem
to constitute a satisfactory application of
this principle.

Figure 20 shows a vertical, longitudi-
nal section, Figure 21 a plan, and Figure
22 a vertical cross-section of a new form

of flush-tank, of which the inside measurements are, a width of one foot eight inches, a length of six feet, and a height of two feet. At a distance of four inches below the top on the long sides, a ledge two inches wide is formed by setting the brickwork of the walls that far back.

This ledge is intended to receive wire-cloth screens twenty-four inches square, shown in Figure 24. The length of the tank may be increased to any number of multiples of two feet in order to obtain the desired capacity. A tank eight feet long would take four screens, a tank ten feet long, five screens, etc. The uniform width is maintained, as the screens are made only in the one size, and as they would be liable to sag if much wider between their supports. The bottom of the tank is so graded as to deliver its contents entirely at the center of the

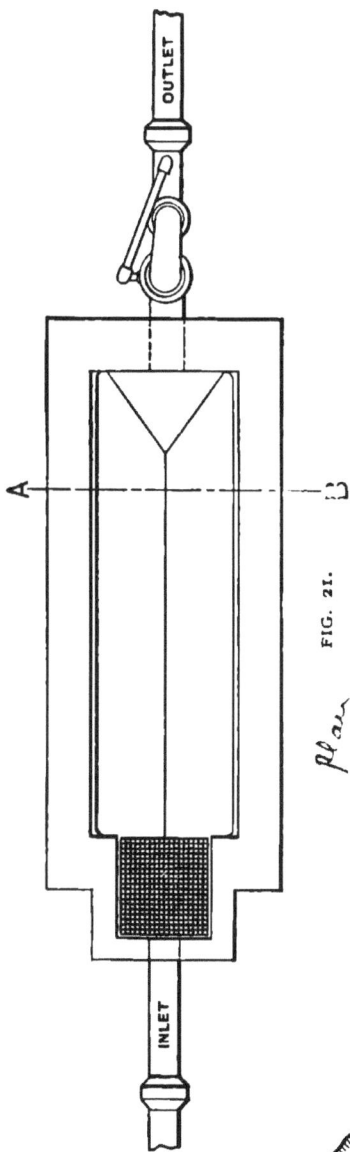

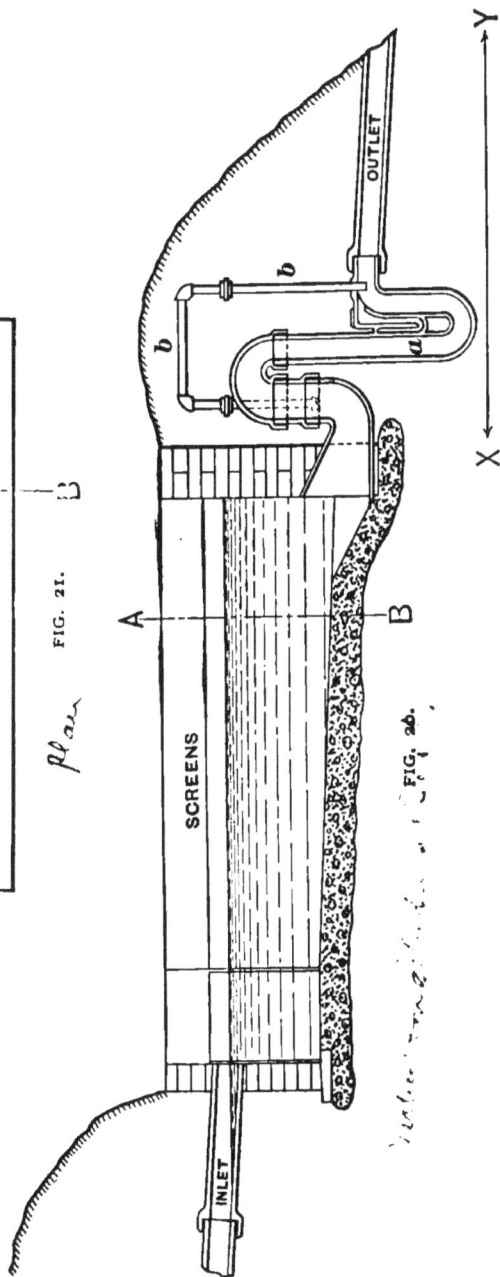

FIG. 21.

Plan

FIG. 22.

OUTLET

INLET

A — B

OUTLET

INLET

SCREENS

A — B

X — Y

discharging-end, where there is a depres-
sion equal to the receiving-height of the
throat of the siphon at its narrowest
part. At the end of the tank there is
built a recess fifteen inches square to
receive the screening-cage shown in

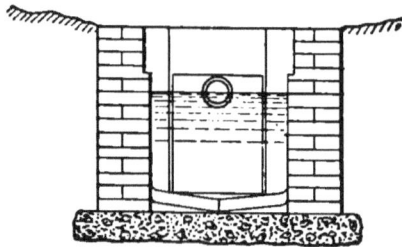

Section A B

FIG. 22.

Figure 23.* This cage is made of
galvanized-iron-wire cloth with 1-inch
meshes. It is entirely closed at the top
and bottom and on three of its sides.
One of its sides, that which is to be
placed next to the inflowing drain, has

---

* As shown in the illustrations, the sewage enters at one end
of the tank and flows out at the other. It will be better to
make the inlet and the outlet at the same end, so that deposits
forming near the inlet will have the full flow to remove them.

an opening at its top ten inches square.
This cage constitutes a complete screen
to withhold whatever will not pass a 1-
inch mesh,—paper and all solids of considerable size. The agitation of its contents by the inflow will break up much
of the softer solid parts of the sewage

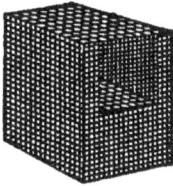 

FIG. 23.   FIG. 24.

and carry them through the meshes;
what will not so pass must be retained,
because it would tend to obstruct pipes
in the case of sub-surface delivery, and
might make objectionable deposits on
the ground in the case of surface
delivery. These cages are furnished in
duplicate, so that whenever one is
removed for cleaning, another can be
substituted for it immediately. The one

removed, after standing a few minutes, will have parted with all of its liquids, and its solid contents can be shaken out through the 10-inch opening and removed, or dug into the ground. When the cage and the covers, Figures 23 and 24, are all in place, the whole tank is sufficiently screened from observation and is protected against leaves and rubbish which might otherwise get access to its contents. As often as experience shows it to be necessary, perhaps daily, the covering-screens, Figure 24, should be removed, after discharging the tank, and its walls and bottom should be thoroughly swept down, the sewage accumulated in its outlet being sufficient for such washing. As above indicated, the frequency with which this cleansing should be performed may vary according to nearness to, or remoteness from, the house, walks, etc.

Figure 29 shows the masonry construction of this tank, the material being brick, glazed on the inner face, and marble or other slabs at the top.

The tank is emptied, after its contents reach a certain height, by the action of a suitable automatic siphon, placed entirely outside of the tank, having a funnel-shaped inlet for the entrance of the sewage. This siphon will require no attention. Whenever the tank fills to the discharging-line, the whole accumulation will flow out rapidly, and when the flow ceases the siphon will "break," allowing no further discharge until the tank has filled again.

The tanks may be built of ordinary brickwork, laid and coated inside and out with Portland cement, or with stone or concrete similarly coated. They may be cheaply but simply made, or they may even be lined and capped with white marble. Another excellent material would be white or straw-colored glazed brick laid with close joints. Elegance of finish will not be altogether useless ; for the finer they are the more easily and the more certainly will these tanks be kept in good condition.

In a few cases, all others so far heard from working satisfactorily, it was found that the flocculent matter passing the screen clogged the 4-inch absorption-tiles after a time. This was obviated by constructing a settling-chamber, such as is shown at the left end of Figure 12, in the course of the drain from the house to the flush-tank. This may be necessary in other cases. For surface disposal, it has been clearly shown not to be required.

Figure 25 shows the flush-tank, illustrated in Figures 20 to 24, placed at some distance from the house, receiving sewage from three house-drains, and delivering its contents for surface disposal by the use of three alternating systems of surface gutters or barriers to

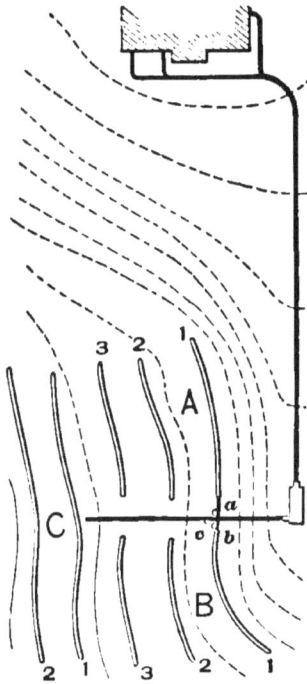

FIG. 25.

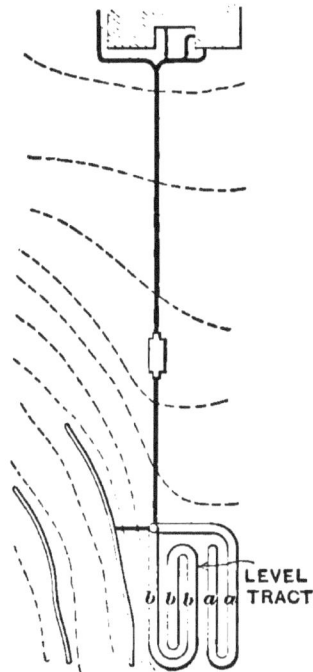

FIG. 26.

equalize the flow. These sections are marked *A*, *B*, and *C*, the gates *a*, *b*, and *c* regulating the distribution. In *A* and *B* the sewage is delivered along the

upper edge of gently sloping land.   If
the land is steeper, the gutters or barriers
must be nearer together to equalize the

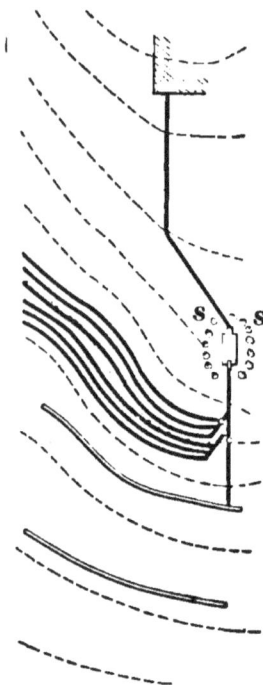

flow.   Water escaping
from the upper gutter,
or barrier 1, is collected
again for a uniform flow
at barrier 2, and again
at barrier 3.   Disposal
for section *B* operates
in the same manner.   On
section *C*, the gutters or
barriers   being   much
longer,   only   two   are
needed.   These illustra-
tions   are   not   drawn   to
scale,   and   are   only   in-

FIG. 27.

tended   to   illustrate   the general features
of the process.   The gutters or barriers
must be absolutely horizontal, and so
arranged that the sewage escaping from
them will flow evenly over the land

below. The distribution may require the
cutting of a leader-furrow here and there
in the grass with a spade.

This method of surface irrigation

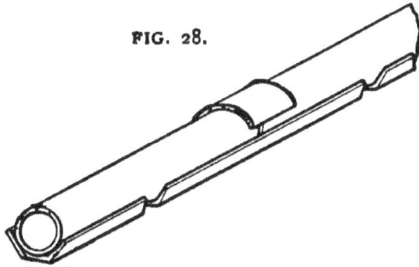

FIG. 28.

removes absolutely all impurity from the
sewage ; what becomes of it after it has
passed over a sufficient area of ground is
immaterial. If it escapes into the brook
or other water-course, it will by that time
have become purer than the water of the
brook itself.

In place of the gates *a*, *b*, and *c*, for
changing the flow from one tract to
another, I am now using a cast-iron distributing chamber, shown in Figure 30.

Its operation is indicated by Figures 31,
32, and 33. The pipe A from the flush-
tank delivers into this, and from its
opposite side run three lines, leading to

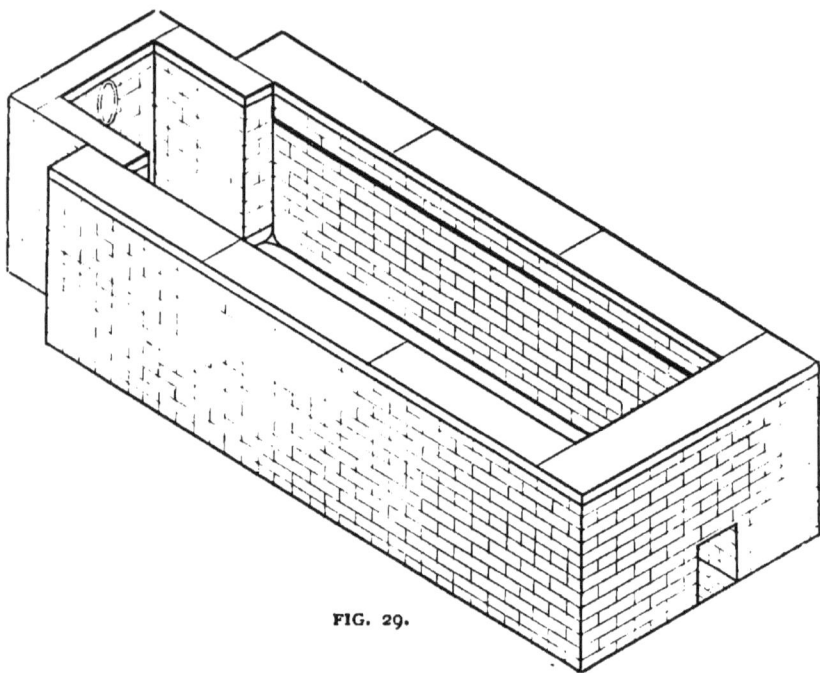

FIG. 29.

three sets of absorption tiles, x, y and z.
Flat plates, which act as deflectors, are
hinged at h, h, fitting closely to the bottom of the chamber. When these plates
are set as shown in Figure 31, the flow is

delivered to line *y*. When turned to the position indicated in Figure 32, the sewage is discharged into *z*. When placed as shown in Figure 33, the outlet *x* receives the flow. These distributing chambers are made by the Dececo Co., 146 High Street, Boston, Mass.

Figure 26 shows a system in which the same tank is used, receiving the flow from four house-drains, and delivering its sewage into absolutely level, wide trenches, of sufficient length. In the case shown, there are two of these trenches returned on themselves to give sufficient length. They are marked *a*, *a*, *a*, and *b*, *b*, *b*. In connection with the same system, there is shown a system of surface irrigation on sloping land. The satisfactory use of the trenches, *a*, *a*, *a*, and *b*, *b*, *b*, requires land of very absorptive character, the more porous the better. The best of all is a very fine gravel.

As the trenches become filled on the discharge of the flush-tank, the liquid soaking away into the ground, there is left
a felt-like coating on the surface, which
requires either a sufficient intermission

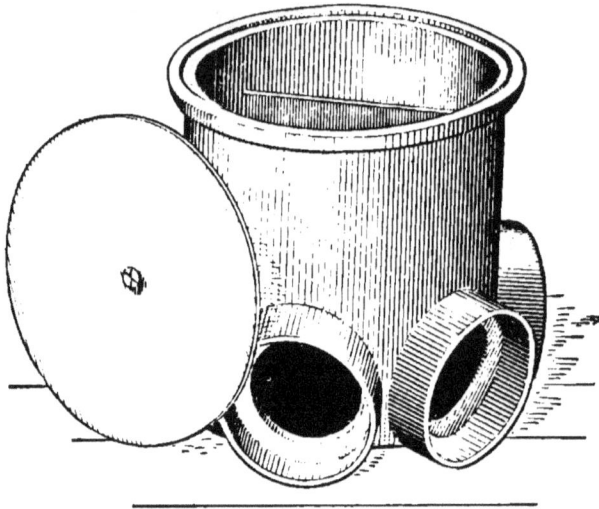

FIG. 30.

of use to be destroyed by exposure to
the air, or, what accomplishes the same
purpose, a thorough raking of the surface from time to time into the material
in which the trenches are cut.

Figure 27 shows the same flush-tank
surrounded by a screen of evergreens,

s, s; two systems of sub-surface absorption-drains similar to those shown in
Figures 14, 15, and 16; and one system
of surface disposal similar to those
shown in Figure 25.

The variation of details, such as the
size and location of the flush-tank, the
arrangement, location, and extent of the
surface gutters, or barriers, the horizontal trenches, the sub-surface absorption-
drains, etc., may be almost infinite, so
that the character of the soil, the forma-
tion of the surface, the use to which the
land is to be put, the necessity for concealment, etc., may be accommodated in
all cases. The flow from the flush-tank
to the absorption-field is conveniently
directed to the different sections by a
simple gate-chamber made for the purpose.

While gutters cut into the ground will
be effective in collecting and equalizing

the flow, they have the drawback that
they retain sewage after the flow ceases,
and become odorous.   The porous barriers (of broken stone, gravel, etc.) allow

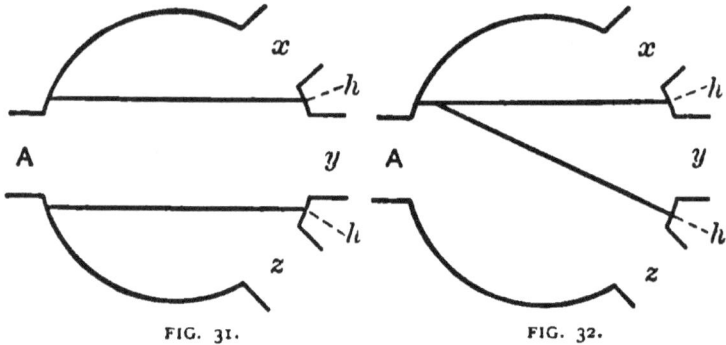

FIG. 31.   FIG. 32.

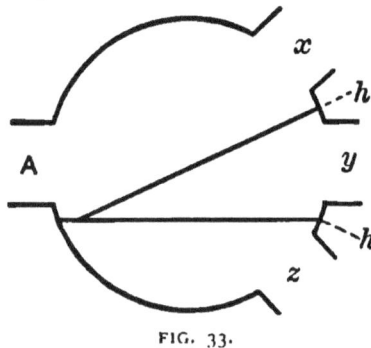

FIG. 33.

the whole flow to pass with only such
delay as is needed to equalize the distribution.   It is best to lay these barriers on a narrow strip of brick paving.
It may with advantage be repeated

here that while this system of disposal seems to be as nearly perfect as is possible in the present state of the art, no such system will withstand neglect. It affords a perfect solution of one of the most difficult and dangerous problems connected with life in districts where sewers are not available, and the completeness of the result to be secured amply compensates for the slight amount of regular attention required.

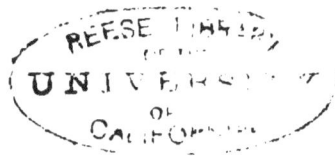

www.ingramcontent.com/pod-product-compliance
Lightning Source LLC
Chambersburg PA
CBHW021656210326
41599CB00013B/1442